Springer Proceedings in Physics 61

Springer Proceedings in Physics

Managing Editor: H. K. V. Lotsch

44 *Optical Fiber Sensors*
Editors: H. J. Arditty, J. P. Dakin, and R. Th. Kersten

45 *Computer Simulation Studies in Condensed Matter Physics II: New Directions*
Editors: D. P. Landau, K. K. Mon, and H.-B. Schüttler

46 *Cellular Automata and Modeling of Complex Physical Systems*
Editors: P. Manneville, N. Boccara, G. Y. Vichniac, and R. Bidaux

47 *Number Theory and Physics*
Editors: J.-M. Luck, P. Moussa, and M. Waldschmidt

48 *Many-Atom Interactions in Solids*
Editors: R .M. Nieminen, M. J. Puska, and M. J. Manninen

49 *Ultrafast Phenomena in Spectroscopy*
Editors: E. Klose and B. Wilhelmi

50 *Magnetic Properties of Low-Dimensional Systems II: New Developments*
Editors: L. M. Falicov, F. Mejía-Lira, and J. L. Morán-López

51 *The Physics and Chemistry of Organic Superconductors*
Editors: G. Saito and S. Kagoshima

52 *Dynamics and Patterns in Complex Fluids: New Aspects of the Physics-Chemistry Interface*
Editors: A. Onuki and K. Kawasaki

53 *Computer Simulation Studies in Condensed Matter Physics III*
Editors: D. P. Landau, K. K. Mon, and H.-B. Schüttler

54 *Polycrystalline Semiconductors II*
Editors: J. H. Werner and H. P. Strunk

55 *Nonlinear Dynamics and Quantum Phenomena in Optical Systems*
Editors: R. Vilaseca and R. Corbalán

56 *Amorphous and Crystalline Silicon Carbide III and Other Group IV – IV Materials*
Editors: G. L. Harris, M. G. Spencer, and C. Y.-W. Yang

57 *Evolutionary Trends in the Physical Sciences*
Editors: M. Suzuki and R. Kubo

58 *New Trends in Nuclear Collective Dynamics*
Editors: Y. Abe, H. Horiuchi, and K. Matsuyanagi

59 *Exotic Atoms in Condensed Matter*
Editors: G. Benedek and H. Schneuwly

60 *The Physics and Chemistry of Oxide Superconductors*
Editors: Y. Iye and H. Yasuoka

61 *Surface X-Ray and Neutron Scattering*
Editors: H. Zabel and I. K. Robinson

62 *Surface Science: Lectures on Basic Concepts and Applications*
Editors: F. A. Ponce and M. Cardona

63 *Coherent Raman Spectroscopy: Recent Advances*
Editors: G. Marowsky and V. V. Smirnov

64 *Superconducting Devices and Their Applications*
Editors: H. Koch and H. Lübbig

Volumes 1–43 are listed on the back inside cover

H. Zabel I. K. Robinson (Eds.)

Surface X-Ray and Neutron Scattering

Proceedings of the 2nd International Conference, Physik Zentrum, Bad Honnef, Fed. Rep. of Germany, June 25–28, 1991

With 120 Figures

Springer-Verlag
Berlin Heidelberg New York
London Paris Tokyo
Hong Kong Barcelona
Budapest

Professor Dr. Hartmut Zabel
Fakultät für Physik und Astronomie, Ruhr-Universität Bochum,
W-4630 Bochum, Fed. Rep. of Germany

Dr. Ian K. Robinson
AT&T Bell Laboratories, Building 510E, Brookhaven National Laboratory,
Upton, NY 11973, USA

Library of Congress Cataloging-in-Publication Data. Surface X-ray and neutron scattering: proceedings of the 2nd international conference, Physik Zentrum, Bad Honnef, Fed. Rep. of Germany, June 25-28, 1991 / editors, H. Zabel and I. K. Robinson. p. cm. – (Springer proceedings in physics; v. 61) Includes bibliographical references and index.
ISBN-13: 978-3-642-77146-0 e-ISBN-13: 978-3-642-77144-6
DOI: 10.1007/ 978-3-642-77144-6

1. Surfaces (Physics)–Optical properties–Congresses. 2. X-rays–Scattering–Congresses. 3. Neutrons–Scattering–Congresses. I. Zabel, H. (Hartmut), 1946 –. II. Robinson, I. K. (Ian K.), 1955–. III. Series. QC173.4.S94S97 1991 530.4'27–dc20 91-38289

Softcover reprint of the hardcover 1st edition 1992

Typesetting: Camera ready by authors

54/3140-543210 – Printed on acid-free paper

Preface

The field of surface X-ray and neutron scattering has grown substantially in the last few years, as witnessed by the increased participation in this conference since the first meeting in Marseille in 1989. This is partly due to increased awareness of the power of the techniques we use, but is also stimulated by the availability of new central facilities, such as synchrotron radiation sources and neutron reflectometers.

This volume represents the combined contributions of the 88 participants. While entirely confined to the subject of the conference, the contributions come from a wide variety of fields, ranging from physics to biology. Some papers concentrate on basic principles of physics, such as the existence of two-dimensional crystals, mathematically and in reality. Several papers on instrumentation, experimental optimisation and techniques of data analysis were presented. This shows the present rapid evolution of the methods themselves. The high level of discussion of the interpretation of results and of the definition of terms used also proved the need for a conference for researchers with these interests. Our field clearly demonstrates its youth; we felt that it would one day reach maturity, the conference series would gracefully come to an end, and the participants would return to their fields of origin. But that time has certainly not yet come. The future outlook for our new field is very bright. We expect increased use of X-ray and neutron diffraction and reflectivity methods, which will be especially promoted by the new facilities under construction. One participant noted, for example, that every "collaborative access team" proposal for the Advanced Photon Source includes some aspect or other of surface diffraction. It is likely that growth will be particularly aggressive in the technologically relevant areas, such as microelectronics, coatings, corrosion, electrochemistry, and drug delivery systems. For this reason our surfaces may be moving out of the special, rarified conditions of ultra high vacuum and on into the real world of general interfaces.

The editors would like to thank the speakers for promptly preparing their manuscripts and returning them in final form after they had undergone a refereeing process, which was mainly undertaken during the conference. We are grateful to the members of the local committee, N. Metoki, A. Stierle, P. Sonntag, and A. Schreyer, for their help in coordinating the paper handling and all other details essential for smooth running of the conference. We would also like to express our gratitude to the staff of the Physikzentrum, Bad Honnef, for their technical help and the Deutsche Forschungsgemeinschaft for their generous financial support. Thanks go also to the sponsors of the conference. We thank all members of the International Advisory Committee for their helpful suggestions and recommendations.

Bochum, Brookhaven
July 1991

Hartmut Zabel
Ian K. Robinson

Contents

Part V Electrochemistry

Part VI Thin Films and Multilayers

Conference Summary

S.K. Sinha

Corporate Research Laboratories,
Exxon Research & Engineering Co., Annandale, NJ 08801, USA

The overall conclusion at which one arrives from this meeting is that the field of X-ray and neutron scattering from surfaces is one of the most active, rapidly growing, and competitive fields in condensed matter and materials science research. This explosion of interest and activity can probably be traced to the increasing exploitation of synchrotron X-ray techniques for surface studies and also to the rapid development of neutron reflectometers at some of the best neutron facilities around the world. In this context it is worth recalling a little history. X-ray specular reflectivity studies were pioneered by Parratt in his classic work almost 40 years ago using a tube X-ray source, but using methods of data collection and analysis that are essentially those in use today. Present X-ray reflectivity studies, especially of liquids, owe much also to the pioneering work of Pershan and Als-Nielsen, and also Stuart Rice and his collaborators in the early eighties. One of the earliest demonstrations of grazing-incidence X-ray diffraction studies of surfaces was carried out by Eisenberger and Marra in 1979, while the concept of truncation rods and detailed structural studies of surfaces using X-rays were driven in the eighties by the work of Ian Robinson and his collaborators in the U.S., and Cowley and co-workers in Edinburgh. Neutron reflectivity also has a venerable history, going back, I believe, to Fermi just after World War II. Workers such as Passell and others at Brookhaven used the technique in the 1960's to measure the scattering lengths of nuclei such as ^{3}He. Specular reflectivity studies of polymer films were carried out by Stamm at Brookhaven in the early 1980's. Much of the current impetus for magnetic and polymer surface studies has come from the successful programs started by Felcher at Argonne and the group led by Thomas and Penfold at the Rutherford CRISP facility. Zabel, Majrczak and others have more recently been among the first to extend the grazing incidence diffraction technique to neutrons. Finally, in the last few years, the role of non-specular or diffuse scattering from surfaces has begun to be explored with increasing interest.

It is clear from the quality of the talks presented that the current scattering experiments are both sophisticated and of excellent quality. What is perhaps most exciting is that we are working on one of the important interfaces (no pun intended) with industry and technology in this endeavor. There is no doubt in my mind that this work will have an increasing impact on areas as diverse as polymer coatings, adhesion, lubrication, thin film and semiconductor devices, composites, corrosion, electrochemistry, catalysis, surface modification, drug delivery, surfactants and membranes, to give only a partial list! We have seen examples of such potential applications in talks presented here.

From the beautiful papers on surface crystallography and structure, I was reassured that we are not, after all, going to be put out of business by other "imaging" microscopies such as STM, AFM, etc. In many important cases, both types of techniques (microscopy <u>and</u> diffraction) are needed to fully nail down a detailed structure, while for surface phase transitions scattering can provide <u>global</u> statistical information more easily than local imaging. It is clear, however, that every "well-dressed" surface scatterer should have access to results of techniques such as STM, AFM, LEED, or atomic scattering as complementary tools.

I will end my remarks with two comments on interesting technical points raised during the meeting. One concerns the issue of the coherence length of the X-ray (or neutron) beam. There is, of course, both a longitudinal coherence length (ξ_L) and a transverse coherence length (ξ_T) associated with such a beam, which may (or may not) be completely defined by the resolution of the instrument used to prepare the beam. The maximum separation L of points on a surface where two beams can interfere with each other is ξ_L or ξ_T divided by $\sin(\alpha)$, where α is the grazing angle of incidence. If one is investigating a heterogeneous surface, e.g., with stepped and planar regions or with islands (patches) of a different thickness covering another surface, the length scale of these inhomogeneities relative to L must determine whether we are to treat the scattering as from one coherent (rough) surface or as an incoherent superposition of scattering from the two types of regions. It is clear that we need to understand ξ_L, ξ_T somewhat better in our experiments. The second remark concerns the dreaded "phase-problem" for inverting observed data to yield structure information. It is important to note that <u>both</u> X-ray reflectivity <u>and</u> standing wave spectroscopy yield phase information in the sense that they involve interference between scattering from the "sample" and some reference surface (e.g., the substrate). However, if the density profile of the "sample" is <u>not</u> symmetric, this information is not enough to yield the complete density profile, which actually involves two functions (the real and imaginary parts of its Fourier transform). Thus, one must vary <u>something</u> (e.g., the substrate reflectivity or its distance to the "sample" or the sample scattering relative to the substrate) and repeat the measurement to get <u>all</u> the information. What one ends up with is a sort of Fourier <u>holography</u> for the sample (in one direction only - normal to the surface).

Finally, it is inappropriate to end without expressing our collective thanks to Ian Robinson, Harmut Zabel, and the local organizing committee for a truly rewarding and excellently run conference.

Part I

Surface Crystallography and Phase Transitions

Surface X-Ray Crystallography and STM Images

R. Feidenhans'l

Physics Department, Risø National Laboratory,
DK-4000 Roskilde, Denmark

STM images have recently shown to be very useful in the interpretation of surface x-ray diffraction data. We will in this paper discuss this issue by giving examples from the Cu(110)c6x2-O and Au/Si(111) 5x1 and 6x6 surfaces.

Introduction

Surface x-ray diffraction has become an important tool in the crystallography of surfaces [1,2]. The strength of the technique is based on the kinematical diffraction of the x-rays, which simplifies the data interpretation and allows the use of standard crystallographic techniques. Direct information about the atomic geometry can be obtained from a contour plot of the Patterson function. However, this only gives a plot of the interatomic vectors, not the electron density itself. A direct atomic view of the surface can be achieved by Scanning Tunnelling Microscopy (STM). Although STM images can not always be interpreted directly as atomic topology, the images can give very important information about interatomic vectors and atomic symmetries. We will show some examples, where we only will discuss the atomic structure *in-plane*, that is only lateral with the surface plane, not perpendicular to it.

Crystallographic data analysis

In the x-ray diffraction experiment, a set of structure factor intensities $\{|F_{hk}|^2\}$ are measured. The structure factors are the Fourier components of the electron density. In general the structure factors are complex and have an amplitude and a phase

$$F_{hk} = |F_{hk}|e^{i\alpha_{hk}}. \quad (1)$$

Unfortunately, only the amplitudes are experimentally available, and hence the electron density cannot be derived directly. Only in very favorable cases where the phases can be deduced from symmetry arguments is it possible to determine the electron density [3]. Very useful information about the atomic geometry can be obtained from a contour plot of the Patterson (autocorrelation) function

$$P(x,y) = \sum_{hk} |F_{hk}|^2 cos(hx + ky). \quad (2)$$

The plot of the Patterson function gives a map of the interatomic vectors between the atoms in the structure [4]. On basis of this map a trial model must be proposed and tested by comparing the measured and calculated structure factor intensities. The Patterson map can be used to exclude models proposed earlier in literature. More difficult , however, can it be to propose a trial model able to explain the features in the map. It is not always necessary to get the true structure right in the first trial. If the main part of the model structure is correct, the remaining parts can eventually be found in an electron density difference plot - a standard trick from bulk x-ray crystallography. The structure factors are the Fourier components of the electron density. If the trial structure is close enough to being correct, the phases of the measured structure factors can be substituted by the phases of the structure factors from the model. Then the electron density difference, the difference between the true structure and the model can be calculated by [4]

$$\Delta\rho(x,y) = \sum_{hk}(|F_{hk}^{true}| - |F_{hk}^{model}|)cos(2\pi(hx+ky) - \alpha_{hk}^{model}). \qquad (3)$$

This will highlight the missing parts of the structure and lead to a completion of the structure.

Clearly, the crucial part in the structural determination is the trial model which must be proposed on basis of the Patterson plot. The model must be close to the true structure, otherwise the electron density difference map will not work. It is at this point where STM images can be very useful. The advantage of STM is that it directly gives a map of the surface. The drawbacks of STM as a crystallographic tool is that the interpretation of the map in terms of atomic geometry not always is obvious and it does not provide any chemical specificity. However, the combination a surface x-ray diffraction and STM images is a very powerful symbiosis for surface crystallography.

Cu(110)c6x2-O

The Cu(110)c6x2-O surface serves as a good example [5]. The unit cell is sufficiently large that only very few models have been proposed in literature. A contour map of the Patterson function in the irreducible unit is shown in fig. 1. Apart from the origin peak, five peaks can be seen in the plot. The x-ray scattering cross section for an atom is proportional to the atomic number Z. For Copper Z=29 and for Oxygen Z=8. Therefore x-ray scattering will mostly be sensitive to the Copper atoms and the position of these have to be determined first. A STM image is also shown in fig. 1. The arrangement of white protrusions in a c6x2 arrangement is clearly seen. Additionally, faint features are seen in between the protrusions. Most of the Patterson plot and the STM-image can be explained by a model consisting only of copper atoms shown in fig. 1. The Copper 1 atom corresponds to the protrusion in the STM-image and the quartet of Copper 2 atoms correspond to the faint features. This model gives a reasonably good agreement between measured and calculated structure factors. Because the Copper atoms constitutes most of the electron density in the surface, the model consisting of only Copper atoms is close to the true electron density and the remaining part - the Oxygen atoms - can be found in a difference plot [5]. The Oxygen atoms are invisible in the STM image.

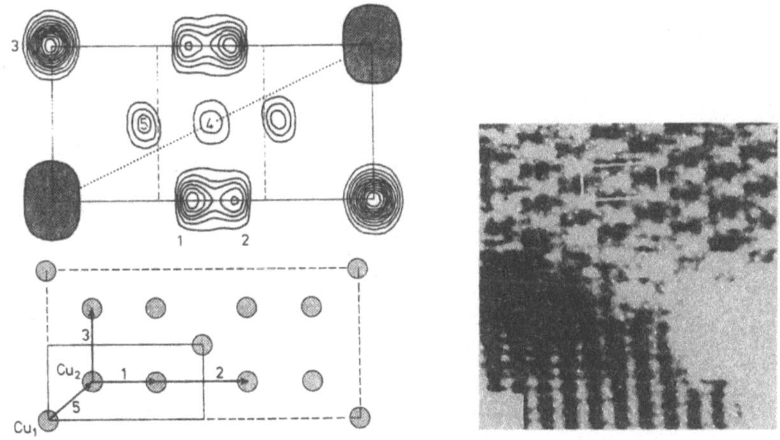

Figure 1: *(Top, left) Contour map of the Patterson function for the Cu(110)c6x2-O Structure. The rectangle corresponds to the solid rectangle in (b). (Bottom, left) A model consisting only of Cu-atoms explaining the main features of the Patterson map in 1a. (Right) The upper half is STM-images from the Cu(110)c6x2-O structure, the lower half is a 2x1 structure. The c6x2 unit cell is indicated.*

Au on Si(111)

The structures of Au monolayers on Si(111) surfaces have been studied with a variety of techniques. Briefly, the system forms a "5x1", a $\sqrt{3}$x$\sqrt{3}$ and finally a 6x6 structure as a function of increasing Au coverage. STM work by Baski, Nogomi and Quate [7,8] showed that the structures were very complicated with a substantial amount of inherent disorder. The STM image for the 5x1 structure is shown in fig. 2. The "5x1" structure consists of disordered rows of 5x2 unit cells giving rise to streaks of scattered intensity at half-order positions [6]. As seen in the STM image the content of the unit cell varies from unit cell to unit cell so the Au atoms have partial occupancies. The STM images show very bright protrusions together with round, rectangular and triangular units, the latter labelled A,B,C, respectively, see fig. 2. It was recently possible to in detail determine the Au/Si(111) structure on basis of x-ray diffraction data [9]. The model is shown in fig. 2 and clearly relates in the atomic details with the STM image.
The 6x6 structure saturating at coverages higher than 1 ML shows a new complication: The structure breaks the symmetry imposed by the Si substrate, which leads a 6-fold twinned structure. This is already evident from the STM image [8] in fig. 3. The long diagonal in the 6x6 cell shown is not a mirror line, as it would have been for a nonc twinned structure. Twinned structures are very difficult to analyze because they share the same points in reciprocal space, which means that the intensities measured are averaged over all domains. Therefore, information from STM images is gratefully acknowledged. The STM images consist mainly of white protrusions. In the x-ray analysis [10] these were

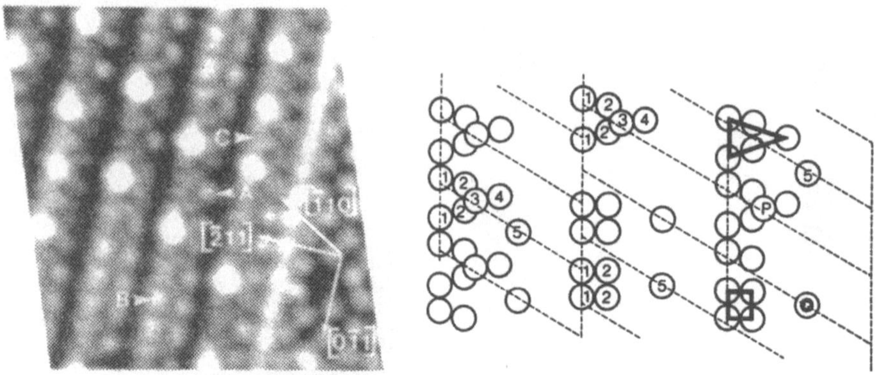

Figure 2: *(Left) STM image of the Au/Si(111)5x1 structure. From ref. 7. (Right) Atomic model of the Au atoms in the 5x2 unit cells.*

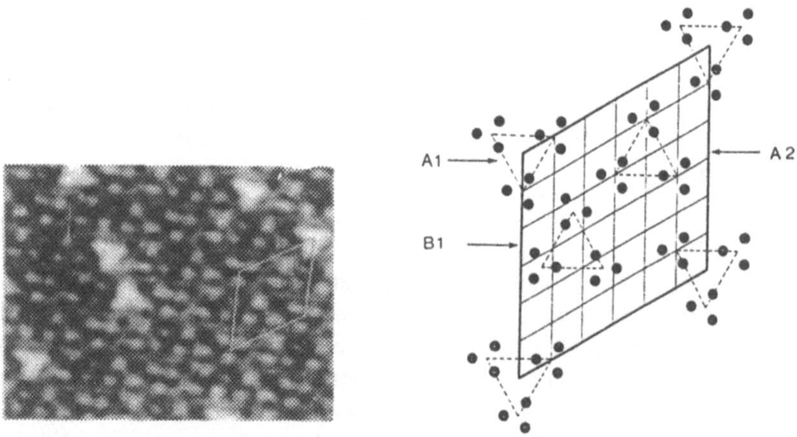

Figure 3: *(Left) STM image of the Au/Si(111)6x6 structure. The unit cell is indicated. From ref. 8 . (Right) Atomic structure determined by x-ray diffraction. The Au-trimers correspond to the white protrusions in fig. 3.*

identified not as individual atoms, but as trimers of Au atoms. The model derived by x-ray diffraction is shown in fig. 3. The coverage of Au atoms in the model is only 0.75 ML, which is below that reported experimentally, see ref [10]. Unfortunately, the twinning of the structure prevents the use of a elecron density difference map to locate the remaining Au atoms.

Conclusion

The combination of x-ray diffraction and STM images can be very powerful for surface crystallography as shown by the examples given in this paper. The STM images can suggest trial models which agree with the Patterson plots and can also suggest including disorder or broken symmetries, which otherwise would be difficult to obtain from the diffraction data alone.

Acknowledgements

This work was supported by the Danish Center for Surface Reactivity. All the x-ray diffraction measurements reported here were performed in collaboration with F.Grey, M.Nielsen and R.L.Johnson. The STM-images of the Cu(110)c6x2-O surface was taken by F.Jensen, F.Besenbacher, I.Stensgaard and E.Lægsgaard. The structure determination of the Si(111)5x1-Au structure was performed by Ch.Shamper, W.Moritz and H.Schulz and of the Si(111)6x6-Au structure by D.Dornisch, W.Moritz and H.Schultz. All are acknowledged for very stimulating and fruitful discussions. Furthermore, I'm very grateful to C.F.Quate for permitting me to use the STM-images of 5x1 and 6x6 structures of Au/Si(111). This work was supported by the Danish Research Councils through the "Center for Surface Reactivity".

References

[1] R.Feidenhans'l. Surf.Sci.Rep.**10**,105(1989).
[2] I.K.Robinson in "Handbook on Synchrotron Radiation", vol III, ed. D.E.Moncton and G.S.Brown (North-Holland, Amsterdam, 1990).
[3] R.Feidenhans'l, F.Grey, R.L.Johnson, S.G.J.Mochrie, J.Bohr and M.Nielsen. Phys.Rev.**B41**,5420(1990).
[4] B.E.Warren. *X-ray Diffraction* (Addison-Wesley, Reading, MA 1969).
[5] R.Feidenhans'l, F.Grey, M.Nielsen, F.Besenbacher, F.Jensen, E.Lægsgaard, I.Stensgaard, K.W.Jakobsen, J.K.Nørskov and R.L.Johnson. Phys.Rev.Lett.**65**,2027(1990).
[6] H.Lipson, K.E.Singer. J.Phys.C**7**,12 (1974).
[7] A.A.Baski, J.Nogami, C.F.Quate. Phys.Rev.**B41**, 10247 (1990).
[8] J.Nogami, A.A.Baski, C.F.Quate. Phys.Rev.Lett.**65**, 1611 (1990).
[9] Ch.Shamper, W.Moritz, H.Schulz, R.Feidenhans'l, M.Nielsen, F.Grey, R.L.Johnson. Phys.Rev.**B43**, 12130 (1991).
[10] D.Dornisch, W.Moritz, H.Schulz, R.Feidenhans'l, M.Nielsen F.Grey, R.L.Johnson. To be published.

Determination of Metal Adsorbed Surfaces by X-Ray Diffraction

T. Takahashi

Institute for Solid State Physics, University of Tokyo,
Roppongi, Minato-ku, Tokyo 106, Japan

Abstract. The structures of the Si(111)$\sqrt{3}$x$\sqrt{3}$-Ag, and -Ga surfaces were studied using X-ray diffraction of both reflection and transmission cases. The three dimensional positions of the adsorbed atoms with respect to the bulk crystal were determined. Oscillatory behavior of the fractional order rod profiles is also discussed.

1. Introduction

X-ray diffraction has become a powerful tool to investigate the surface structure. In order to obtain the three dimensional surface structure, the measurements of rod profiles are required. For this purpose, we observed the intensity versus energy curves as in the LEED experiments both in the reflection and transmission cases. The experiments were made at the beamline 10C in the Photon Factory. Hitherto we have studied the Si(111)$\sqrt{3}$x$\sqrt{3}$-Bi and -Ag surfaces by this method [1-3].

In this work, we study oscillatory behavior of the fractional order rod profiles in the Si(111)$\sqrt{3}$x$\sqrt{3}$-Ag surface, and also show that Ga atoms in the Si(111)$\sqrt{3}$x$\sqrt{3}$-Ga surface occupy the T_4 sites.

2. Si(111)$\sqrt{3}$x$\sqrt{3}$-Ag Surface

2.1 Outline of Si(111)$\sqrt{3}$x$\sqrt{3}$-Ag Structure

In the Si(111)$\sqrt{3}$x$\sqrt{3}$-Ag surface, there has been controversy as to the points of 1) the top layer atom (Ag or Si), 2) the coverage of Ag atoms (2/3 or 1 monolayer) and their arrangement, 3) the adsorption site with respect to the bulk crystal, and 4) the displacements of Si atoms from their bulk positions. In a previous work [1-3], we determined the first three points by measuring ten integral order and seven fractional order rod profiles. The result is shown in Fig.1.

The top layer is occupied by Ag atoms. Triangles made of three Ag atoms are chained in the honeycomb arrangement as shown in Fig.2. Therefore this model was named the honeycomb-chained triangle (HCT) model. This model differs from the usual trimer model in the point that each Ag atom has four nearest neighbor Ag atoms. It should be also pointed out that the interatomic distance between the Ag atoms is nearly equal to the van der Waals distance, 3.44Å.

The adsorption site with respect to the bulk crystal was determined from the ten integral order rod profiles. In order

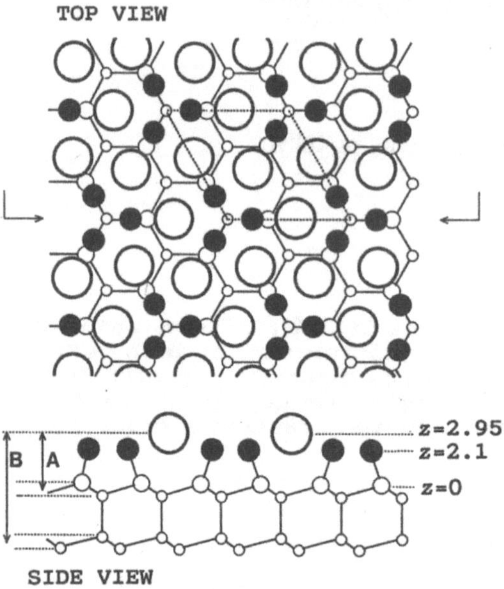

Fig.1 Illustration of the Si(111)$\sqrt{3}$x$\sqrt{3}$-Ag structure. The positions of the Ag atoms (large circles) and the missing top layer Si atoms (solid circles) with respect to the unreconstructed Si crystal are shown. The unit cell is shown by dotted lines.

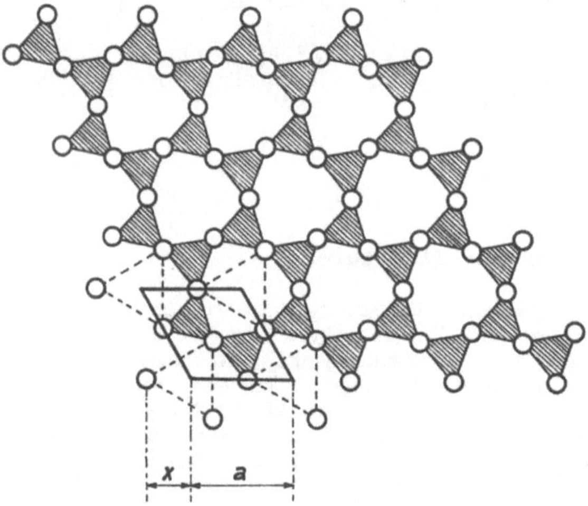

Fig.2 Arrangement of Ag atoms in the Si(111)$\sqrt{3}$x$\sqrt{3}$-Ag structure. Ag triangles are chained in the honeycomb arrangement. The centers of the hatched triangles coincide with the positions of the honeycomb protrusions observed in the STM image [16].

to get high resolution of the lateral adsorption site, transmission X-ray diffraction was applied to the measurement of the (0I) rod profile. The detailed analysis about the reconstruction of the Si crystal indicated that the missing top layer model as shown in Fig.1 is most favorable. The height of the Ag layer from the top of the unreconstructed double layer *z* is (2.95±0.1)Å, and the parameter *x/a* indicating the size of the trimer as shown in Fig.2 is 0.44±0.015, where a is the dimension of the unit cell. For the first Si layer forming the trimer, *z*=(2.1±0.2)Å and *x/a*=0.20±0.03.

2.2 Intensity Oscillations in Fractional Order Rod Profiles

As to the above mentioned three points, similar results were obtained by other works [4-9]. However, the displacements of Si atoms from their bulk positions are yet to be determined. The feature of the rod profiles observed in the Si(111)$\sqrt{3}$x$\sqrt{3}$-Ag surface is that the diffracted intensities of the fractional order rods strongly oscillate, for instance, as seen in Fig.3, which is in contrast to the case of the Si(111)$\sqrt{3}$x$\sqrt{3}$-Bi surface [10].

Since the X-ray scattering from the Ag layer is dominant in this system, the intensity oscillations arise from mainly the interference effect between the X-rays diffracted from the Ag layer and the reconstructed Si layers. The period of the oscillations is almost inversely proportional to the distance between the Ag layer and the reconstructed Si layer. The analysis indicates that at least two kinds of oscillation periods exist as indicated by the arrows A and B, which correspond to the spacings A and B shown in Fig.1. In order to explain the peak B around 1.65Å in Fig.3, the interference effect between the layers with the spacing B indicated in Fig.1 must be taken into account.

If, however, we attribute the origin of the oscillations all to the displacements of the substrate Si atoms, the substrate structure obtained by the least squares method is unrealistic, and it is also difficult to explain the relatively small intensity oscillations in the integral order rod profiles. One possibility to overcome these difficulties is to take into account the effect of the step. The interference effect

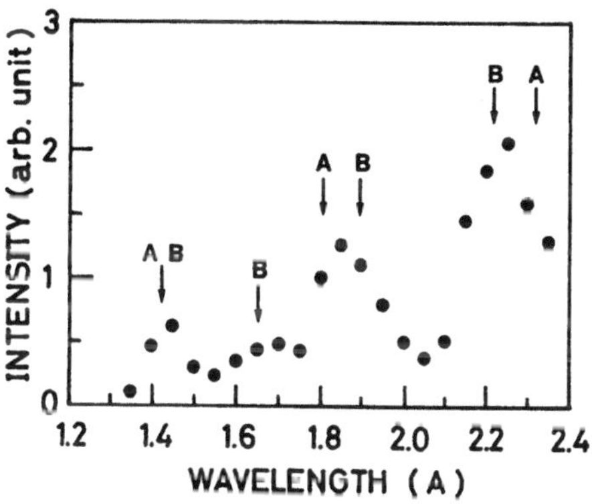

Fig.3 Diffracted intensities along the (2/3 2/3) rod observed for the Si(111)$\sqrt{3}$x$\sqrt{3}$-Ag surface. At least two kinds of oscillation periods exist as indicated by arrows A and B.

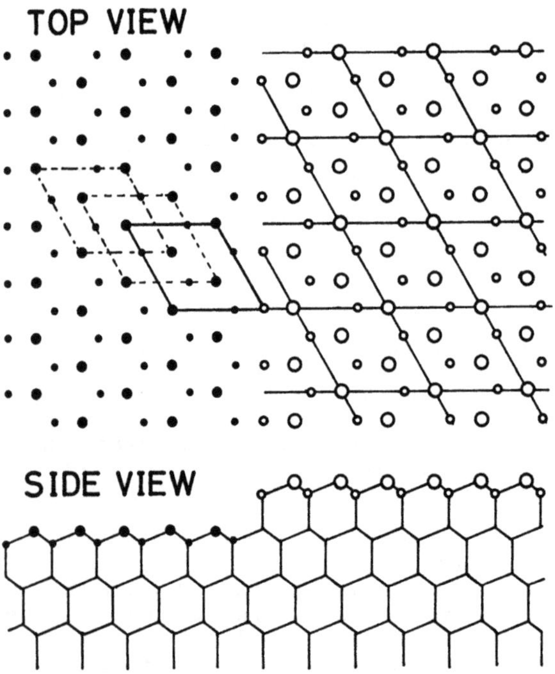

Fig.4 Illustration of three phase relationships between the unit cells at the step.

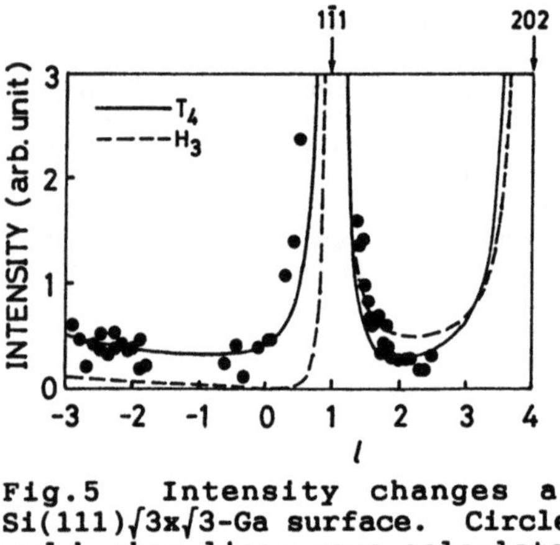

Fig.5 Intensity changes along the (0$\bar{1}$) rod in the Si(111)$\sqrt{3}$x$\sqrt{3}$-Ga surface. Circles are observed points. Solid and broken lines were calculated for the T_4 an H_3 sites, respectively. Arrows indicate the Bragg points.

between the adjacent terraces of the step also gives similar oscillatory behavior in the rod profiles.

In the case of $\sqrt{3}$x$\sqrt{3}$ structure, there exist three phase relationships between the unit cells on different terraces at the step as shown in Fig.4. If strong correlation exists between the structures of the adjacent terraces and only one of the three phase relationships is selectively favored, the above mentioned problems are partly resolved. The analysis showed that the phase relationship between the unit cells indicated by solid lines in Fig.4 is favored. This may mean that the Si(111)$\sqrt{3}$x$\sqrt{3}$-Ag structure grows at the step edge.

3. Si(111)$\sqrt{3}$x$\sqrt{3}$-Ga Surface

The structure of the Si(111)$\sqrt{3}$x$\sqrt{3}$-Ga surface was also studied by X-ray diffraction. From the analysis of the (00) rod profile, the height of Ga atoms was determined to be (1.55±0.05)Å measured from the top of the unreconstructed double layer.

Next, in order to determine the lateral adsorption site, the (0$\bar{1}$)rod profile was measured as seen in Fig.5 using transmission X-ray diffraction. The analysis of asymmetry around the Bragg points shows that only the T_4 site is acceptable and the H_3 site is ruled out. The result obtained in this work is consistent with that in other works [11-13].

4. Discussion

The Si(111)$\sqrt{3}$x$\sqrt{3}$-Ag surface has been studied by STM [14-16]. The model shown in Fig.1 well explains the STM images both in pattern and registration if the protrusions observed in the STM images [16] correspond to the hatched triangles in Fig.2. Recent total energy calculations based on first principles [9] support this interpretation. The result that the interatomic distance between Ag atoms is nearly equal to the van der Waals distance also seems to support this.

The analysis of rod profiles requires enormous calculations depending on models. In principle, however, the Fourier analysis of the rod profiles has some information of the surface structure although the surface structure has no periodicity in the direction of surface normal. The development of the Fourier analysis method is expected to attain the model independent analysis.

In summary the effect of the step on the rod profile is discussed in the $\sqrt{3}$x$\sqrt{3}$-Ag surface, and the adsorption site of Ga in the $\sqrt{3}$x$\sqrt{3}$-Ga surface is determined to be T_4 site.

References

[1] T. Takahashi, S. Nakatani, N. Okamoto, T. Ishikawa and S. Kikuta: Jpn. J. Appl. Phys. **27** (1988) L753

[2] T. Takahashi, S. Nakatani, N. Okamoto, T. Ishikawa and S. Kikuta: Rev. Sci. Instrum. **60** (1989) 2365.

[3] T. Takahashi, S. Nakatani, N. Okamoto, T. Ishikawa and S. Kikuta: Surf. Sci. **242** (1991) 54.

[4] A. Ichimiya, S. Kohmoto, T. Fujii and Y. Horio: Appl. Surf. Sci. **41/42** (1989) 82.

[5] E. Vlieg, A. W. Denier van der Gon, J. F. van der Veen, J. E. Macdonald and C. Norris; Surf. Sci. **209** (1989) 100.

[6] E. Vlieg, E. Fontes and J.R. Patel: Phys. Rev. **B43** (1991) 7185.

[7] I. Kamiya, M. Katayama, E. Nomura and M. Aono: Surf. Sci. **242** (1991) 404.

[8] R.S. Daley, R.M. Charatan and R.S. Williams: Surf. Sci. **240** (1990) 136.

[9] Y.G. Ding and C.T. Chan and K.M. Ho: in preprint.

[10] T. Takahashi, S. Nakatani, T. Ishikawa and S. Kikuta: Surf. Sci. **191** (1987) L825. The result of the fractional order rod profile is unpublished.
[11] A. Kawazu and H. Sakama: Phys. Rev. **B37** (1988) 2704.
[12] J. Nogami, S. Park and C.F. Quate: Surf. Sci. **203** (1988) L631.
[13] J. Zegenhagen, J.R. Patel, P. Freeland, D.M. Chen, J.A. Golovchenko, P. Bedrossian and J.E. Northrup: Phys. Rev. **B39** (1989) 1298.
[14] R. J. Wilson and S. Chiang: Phys. Rev. Lett. **58** (1987) 369.
[15] E. J. van Loenen, J. E. Demuth, R. M. Tromp, and R. J. Hamers: Phys. Rev. Lett. **58** (1987) 373.
[16] R. J. Wilson and S. Chiang: Phys. Rev. Lett. **59** (1987) 2329.

Au Adsorption on Si(111) Studied by Grazing Incidence X-Ray Diffraction

C. Schamper[1], *D. Dornisch*[1], *W. Moritz*[1], *H. Schulz*[1], *R. Feidenhans'l*[2], *M. Nielsen*[2], *F. Grey*[3], *and R.L. Johnson*[4]

[1]Institute of Crystallography and Mineralogy, University of Munich, Theresienstr. 41, W-8000 München 2, Fed.Rep.of Germany
[2]Risø National Laboratory, DK-4000 Roskilde, Denmark
[3]Max Planck Institute for Solid State Research, Heisenbergstr. 1, W-7000 Stuttgart 80, Fed.Rep.of Germany
[4]2nd Institute for Experimental Physics, Hamburg University, W-2000 Hamburg 50, Fed.Rep.of Germany

Abstract. The atomic arrangement in the three ordered structures of Au on Si(111) has been studied by in plane x-ray diffraction. The common feature in all three structures is the formation of small Au clusters and substantial distortions of the substrate. The (5x1) structure consists of a disordered (5x2) structure with partially occupied rows of Au clusters in low symmetric adsorption sites. For the (√3x√3) structure the trimer model could be confirmed. The (6x6) structure consists of trimer triplets located at different adsorption sites and is probably connected with a reconstruction of the top substrate layer.

In the range of monolayer coverages Au forms three ordered structures on Si(111). Between 0.2 and 0.8 monolayers (ML) a (5x1) structure is observed [1]. In agreement with the common usage in the literature the notation (5x1) is used here though this structure is in fact a disordered (5x2) structure as has been noticed already in earlier LEED studies [2]. Above 0.5 ML the (√3x√3)R30° structure occurs and above 1.5 ML the (6x6) structure exists [1]. 1 ML refers to one Au atom per (1x1) substrate unit cell. These structures have been subject of a number of studies with a variety of methods but the atomic arrangement has not become clear until now. Particularly, for the (√3x√3) structure two different structure models have been proposed. A honeycomb structure of Au occupying high symmetric adsorption sites at a coverage of 2/3 [3] and trimer model corresponding to 1 ML coverage [4]. The (√3x√3)R30° structure is also found in the related systems Ag/Si(111) [5] and Ag/Ge(111) [6] and the structure model presented here is closely related to these structures.

We present the structure models as determined from in-plane x-ray diffraction data. The details of the analysis are described in ref. [7,8]. All structure models are in agreement with recent scanning tunneling microscopy (STM) topographs [9-11]. Au was deposited on the clean Si(111)(7x7) surface at a rate of 0.3 ML/min and at a substrate temperature of 400° C. After preparation the sample was transferred under UHV conditions to a small x-ray diffraction chamber. X-ray diffraction data were measured at the wiggler beam line at the HASYLAB [12].

For the (5x1) structure 52 fractional order in plane structure factors were measured. Since the maxima of the patterson function do not agree with Si-Si distances Au atoms adsorb in low symmetric positions [7,12]. The Au sites 1-5 (fig. 1) follow directly from a Patterson map however, elongations of the Patterson

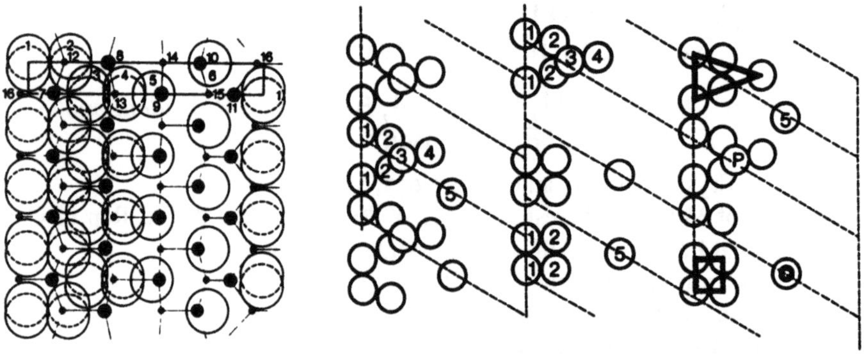

Fig. 1: Au- (large circles) and Si-positions in the first (large dots) and second (small dots) layer as obtained from the refinement of the (5x1) structure (left panel). A model taking account of a (5x2) unit cell and the partial occupation of the Au sites (right panel) explains protrutions (P), round, rectangular and triangular configurations as are observed in (STM) topographs [9].

maxima indicate a disorder in the [110] direction, which can be described by split positions corresponding to a statistical occupation of two different sites. Assuming the symmetry c1m1 the refinement of 25 free parameters, including 6 Au- and 10 Si positions, 5 occupation factors and an overall temperature factor gives a X^2 of 1.7 [7]. The distance of 2.9 Å between the Au sites 1 and 2 agrees with the nearest neighbour distance in the bulk Au. The split positions correspond to the superposition of the (5x2) unit cells into a single (5x1) unit cell. Two distances between Au atoms occur along the rows 1 and 2: 3.10 Å and 4.58 Å. The smaller value comes close to the bulk value of Au. In agreement with STM topographs [9] we find an irregular sequence of small Au clusters along the [110] direction (fig. 1).

14 fractional order structure intensities are used to elaborate the ($\surd$3x$\surd$3) structure including a substrate distortion (fig. 2). Only the trimer model [4] is consistent with the Patterson map as has been reported earlier [8]. The interatomic Au-Au distance of 2.7 Å is slightly smaller than the bulk value. Au Atoms are located near H3 sites, assuming that Si atoms of the top layer are shifted. The registry with the bulk remains uncertain because only superstructure reflections were evaluated. The deviation between the coverage for the ($\surd$3x$\surd$3)-structure found in experiments [1] and the coverage for the trimer model (1 ML) probably results from missing trimers.

The Patterson function for the (6x6) structure was calculated from 139 fractional order intensities (fig. 3). Two main features of the Patterson function have to be discussed. First, it is evident that all interatomic vectors of the ($\surd$3x$\surd$3) structure appear again in the (6x6) structure (peaks 1-3 in fig. 3). Consquently Au trimers remain the structural element of the (6x6) reconstruction. Secondly no peaks in the middle of the unit cell are observed (indicated by A and B in fig. 3). Therefore the double of a translation vector of the ($\surd$3x$\surd$3) unit cell in all directions is missing. This means that the (6x6) structure cannot be explained by a simple superstructure of the ($\surd$3x$\surd$3) reconstruction. Only an arrangement of trimers where the occurrence of three Au atoms in a row is avoided can explain the Patterson function. The structure (fig. 3) has a low symmetry leading necessarily to the existence of 6 twin domains [8]. The coverage of 0.75 ML is well below the experimentally observed coverage of 1

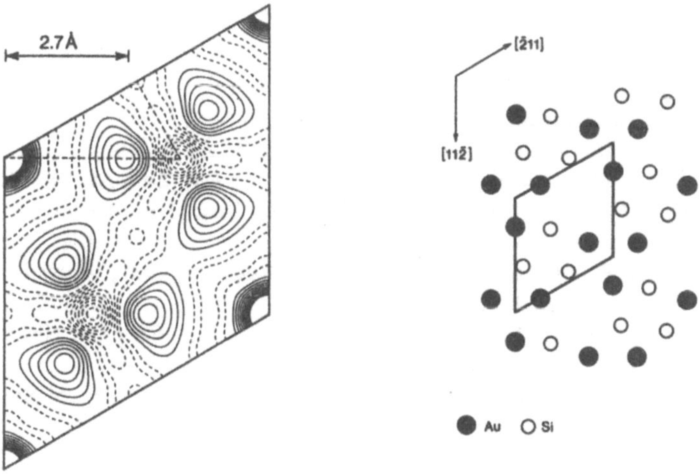

Fig. 2: Patterson function of the (√3x√3) structure (left panel). The irreducible unit cell is marked by dashed lines. In the right panel the trimer model with the 1st distorted Si layer is outlined. H3-like adsorption sites for the Au atoms are assumed.

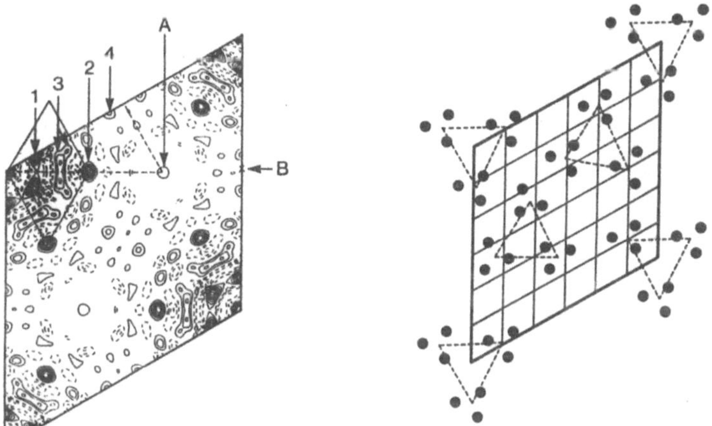

Fig. 3: Patterson function for the (6x6) superstructure (left panel). The shaded area corresponds to a (√3x√3) unit cell. The right panel illustrates a single twin domain of the (6x6) structure.

ML or above. We explain the discrepancy by additional Au atoms either adsorbed in sites between the trimers or diffused into subsurface sites. These Au sites could not be identified in the patterson map and have therefore been left out in the analysis.

Acknowlegement. This work was supported by the Bundesminister für Forschung und Technologie Grant No. 05390CAB and No. 05464IAB8 and the Danish National Research Council. We thank Prof. D. Wolf and Dr. H.L. Meyerheim for useful discussions and the staff of HASYLAB for their help.

References

[1] Le Lay, Surface Sci. 132 (1983) 169
[2] H. Lipson, K.E. Singer, J. Phys. C7 (1974) 12
[3] J.H. Huang, R.S. Williams; Phys. Rev. B83, 4022 (1988)
[4] K. Oura, M. Katayama, F. Shoji, T. Hanawa; Phys Rev. Lett. 55, 1486 (1985)
[5] E. Vlieg, A.W. Denier van der Gon, J.F. van der Veen, J.E. Macdonald, C. Norris, Surf. Sci. 209, 100 (1989)
[6] D. Dornisch, W. Moritz, H. Schulz, R. Feidenhans'l, F. Grey, M. Nielsen, and R.L. Johnson, to be published.
[7] C. Schamper, W. Moritz, H. Schulz, R. Feidenhans'l, F. Grey, M. Nielsen, and R.L. Johnson, Phys. Rev. B 43, 12130 (1991)
[8] D. Dornisch, W. Moritz, H. Schulz, R. Feidenhans'l, F. Grey, M. Nielsen, and R.L. Johnson, to be published.
[9] A.A. Baksi, J. Nogami, C.F. Quate, Phys. Ref. B, 41, 10247 (1990)
[10] T. Hasegawa, K. Hosaka, S. Hosoki, J. Vac. Sci. Technol. A 8 (1) (1990)
[11] J. Nogami, A.A. Baski, C.F. Quate; Phys. Ref. B83, 4022 (1988)
[12] R. Feidenhans'l, F. Grey, M. Nielsen, R. Johnson; in 'Kinetics of Ordering and Growth at Surfaces', Edited by M.G. Lagally, Plenum Press, New York, 1990

Grazing Incidence X-Ray Scattering Study of Staircases of Steps on Si(001) Surfaces

G. Renaud[1], *P.H. Fuoss*[2], *J. Bevk*[2], *and B.S. Freer*[2]

[1]CEN-G/DRFMC/SP2M/S, BP 85X, F-38041 Grenoble Cedex, France
[2]AT&T Bell Laboratories, Murray Hill, NJ 07974, USA

Abstract. Grazing incidence x-ray scattering is used to investigate the regularly spaced steps that result from molecular beam epitaxy of silicon on Si(001) surfaces with a small residual miscut. Precise statistical information on the steps are derived, including the orientation, the average separation and the terrace width disorder. This is applied to study the evolution of the steps distribution as a function of the thickness of the grown overlayer.

1. Introduction

Steps are known to strongly affect the fundamental properties of surfaces, such as electronic surface states and surface reconstructions. They also play a dominant role in the process of thin film epitaxial growth, for instance the realization of quantum wires. Step ordering is known to be strongly affected by Molecular Beam Epitaxy (MBE). In particular, some theoretical calculations show that a decrease of step disorder is expected during homoepitaxy on surfaces with an existing staircase of steps [1]. In this work, Grazing Incidence X-ray Scattering (GIXS) is used to analyze the terrace widths and terrace width disorder on Si(001), and their evolution as a function of the thickness of the grown overlayer during homoepitaxy.

2. Experimental

An initial 4" silicon wafer was cut parallel to the (001) plane within ~0.03°, and polished by a free-floating polishing process, where both sides of the wafer are polished simultaneously to a final roughness of ~1Å [2]. A buffer layer of 2000Å of silicon was first grown on the sample at 1073 K, followed by an anneal for 20 min at 1223 K. A screen was then used to hide successive quarters of the wafer from further MBE deposition. The wafer was then divided into four parts: one without post-anneal deposition, and three portions with an additional 20Å, 500Å and 10000Å of silicon. Step-flow growth conditions were selected, as shown by steady RHEED intensities. A thin layer (typically 10 Å thick) of native oxide formed on the different samples before loading them into the ultra-high vacuum diffractometer. GIXS experiments were performed under grazing incident and exit angles with respect to the surface, using the X16A beamline at the NSLS. The setup of the five-circle diffractometer [3] was kept constant during the study of the four pieces, thus allowing direct comparison of measured intensities.

3. Results

All data have been obtained at {110}, {310} and {200} in-plane positions in the reciprocal space, where the intensity arises because of the sharp truncation of the crystal lattice by a flat surface, which yields rods of intensity extending perpendicular to the macroscopic surface: the so-called Crystal Truncation Rods (CTRs) [4]. Standard analysis of the integrated intensity variation along the rods confirmed that the surfaces were atomically flat. In addition, the magnitude of the integrated intensities of identical reflections were equal to within 20% for the four pieces, which confirmed that there was no significant roughening during deposition.

Fig. 1 shows the intensity measured at the (110) and (200) positions for the four pieces. All CTRs measured at similar positions, for example {110}, {310} and {200}, are similarly split. It was demonstrated [5] that this splitting shows the existence of a regular staircase of steps, arising from a small misorientation of the surface with respect to the (001) atomic planes. For the first piece (no post-anneal deposition), a detailed analysis of the peak splitting values measured at different locations enabled to deduced with a high accuracy the orientation (nearly along the [100] direction) and the value (0.031°) of the residual miscut.

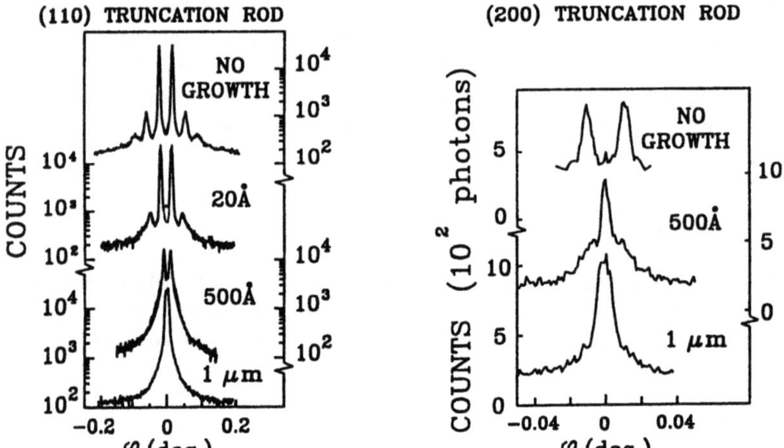

Figure 1: Experimental transverse scan data (thick lines) around the (110) position (left) and the (200) position (right), as a function of the thickness of the MBE grown silicon overlayer. The logarithm of the intensity is plotted as a function of the azimuthal rotation angle ϕ. Also shown (thin lines) are the calculated (110) intensities assuming a Gaussian distribution of terrace widths.[5]

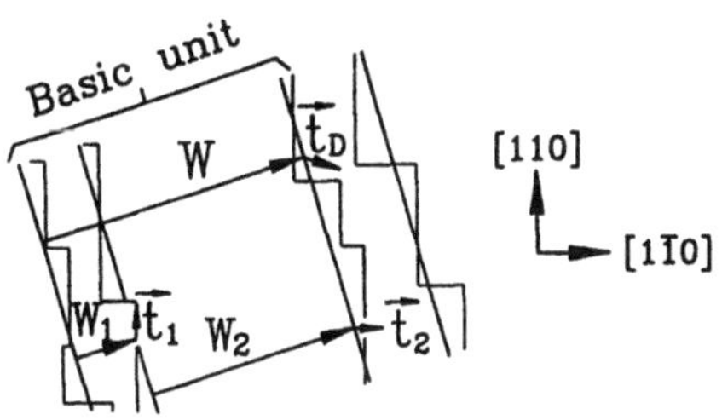

Figure 2. Plane view representation of the bi-terrace constituting the basic unit of the staircase of steps. When crossing successive monolayer high steps, two kinds of translations from the upper terrace to the lower terrace lattices alternate. The mean width of the respective terraces are W_1 and W_2, with a total width of W. On the atomic scale, steps are made of two kinds of different sub-steps, along either the [110] or the [1$\bar{1}$0] directions, separated by kinks; but the overall translation when crossing a step is independent on the detailed atomic structure.

To describe precisely the measured integer order CTR intensity, it is necessary to take into account the discrete nature of the array of steps underlying the miscut surface. The calculations are described elsewhere [5]; we only recall here the main steps. The interference between successive units, giving rise to the observed intensity, arise from the phaseshifts between the waves scattered by successive terraces. These phaseshifts are due to the non-integer translations between the lattices of successive terraces. For the present in-plane scattering configuration, the problem can be reduced to the diffraction by a succession of scattering units made of two adjacent terraces, see Fig. 2. Two kinds of monolayer high steps have to be considered on the Si(001) surface, corresponding to two different in-plane translations between the lattice of the upper terrace and the lattice of the lower terrace: $\vec{t}_1 = 1/4[11\bar{1}]$ and $\vec{t}_2 = 1/4[1\bar{1}\bar{1}]$. The phaseshift between successive units is introduced by the total translation: $\vec{t}_D = \vec{t}_1 + \vec{t}_2 = 1/2[10\bar{2}]$.

At exact integer-order in-plane locations of {110} type, successive bi-terrace units scatter out-of phase, while the two sub-terraces of a unit scatter in-phase. The peak separation is inversely proportional to the whole bi-terrace unit width W, whatever the ratio of the two sub-terraces mean width, W_1/W_2. Expressions were given [5] to simulate experimental data with the introduction of the disorder arising from the different bi-terrace unit widths, but in which the disorder within these units was neglected. They were shown to be good approximations of the {110} and {310} data, regardless of the precise ratio of the two sub-terrace widths and the step disorder. On the other hand, at exact integer-order in-plane locations of {200} type, successive bi-terrace units scatter in phase, while the two sub-terraces of a unit scatter out-of phase. Thus, {200} data can be used to estimate the ratio of the two sub-terraces mean width within the mean bi-terrace unit. In the ideal case of prefect order, it can be shown that if {200} data are perfectly split, the two mean sub-terrace widths are equal: $W_1=W_2$, whereas if there is no splitting, one type of sub-terrace must be narrow with respect to the other: $W_1<<W_2$. The first case is illustrated by the data: after annealing, the (200) CTR is nearly completely split, thus implying that all steps are equally spaced and of monolayer height.

On the starting surface, the staircase of steps is very well ordered, since the (110) diffraction peaks are very narrow, and three orders of diffraction are present and well separated. As the MBE silicon layer thickness increases, there is i) a decrease of diffraction peaks separations, which corresponds to an increase of the mean unit width, ii) an increase of the diffuse component intensity, and iii) a decrease of the number of diffraction orders, which means an increase of the terrace width disorder. All data are well simulated (see Fig.1) by a model [5] based on a Gaussian distribution of terrace widths around the average terrace width, with statistical independence between neighboring terrace widths. On the initial surface, the unit width is W~5000Å, and the rms width variation is σ~670Å, that is 13% of the average mean width. Hence, the two sub-terraces are of equal mean width (as shown above) of ~2500Å, and separated by steps of monolayer height. This very good step ordering is presumably due to dominant evaporation conditions during high temperature annealing [6]. After 20 Å of growth, the mean unit width has risen up to ~6000Å, and the rms variation σ~1100Å, that is 18% of the mean unit width. After 500Å of growth, the mean width W~9300Å has nearly doubled, and the rms disorder has risen up to σ~4000Å, or 43% of the mean width. After 10000Å of growth, the data are no longer split, which shows that terraces are extremely wide on this last sample. The data can be accurately fit with the sum of a Gaussian distribution of 0.010° FWHM (sharp component) and a Lorentzian distribution of 0.034° FWHM (wide component). The width of the sharp component yields an estimation of the pair correlation extension within these terraces, with a characteristic length of 14000Å. The wider component corresponds to an exponentially decaying correlation function, with characteristic length ~4100 Å, which may be attributed to terrace width and step edge disorder.

The question now arises to whether the observed terrace widening during homoepitaxy is due to a uniform decrease of the miscut, or if it is due to a modification of the staircase with higher steps and wider terraces while keeping the miscut constant and uniform, or whether bunches of steps form, with remaining localized regions of lower "local miscut", but still well ordered staircases of steps. The first explanation can be ruled out by simple calculations which show that MBE growth can not modify the miscut uniformly over the whole sampled area. The second mechanism would yield ~4 monolayer high steps after 5000 Å of growth, since the mean unit width has nearly doubled from the original bi-terrace unit width of 5000 Å. In that case, successive terraces would scatter in-phase at the exact {100} positions, and not out-of phase as is experimentally observed. The most likely explanation is thus that, under the chosen growth conditions, bunches of steps form, which do not scatter at in-plane integer order positions. Regions of ordered steps would remain, with a decreasing step density as the steps migrate to the bunches. Variations of the "local miscut" between these different regions is another way to explain the large diffuse scattering after 500 Å and 10000 Å of growth. However, the remaining peak splitting shows that these variations are small.

It would be interesting to know whether steps remain of monolayer height or if they group by pairs as bilayer high steps during this process. It would be tempting to interpret the presence of a strong central contribution at {200} positions after 500 Å and 10000 Å of growth as an evidence for bilayer high steps.
However, two shoulders are present after 500 Å, with a spacing consistent with the width increase determined above. If the terrace width disorder was small, these (200) data could be straightforwardly interpreted as arising from a strongly asymmetric bi-terrace unit, but the observed strong unit width disorder could also be responsible for the central contribution.

4. Conclusions

In conclusion, this study examines the statistical nature of a static array of steps on Si(001) surface after a high temperature anneal and subsequent growth. The high temperature anneal leads to a very well ordered staircase of monolayer high steps: the root mean square terrace width variation is only 13% of the average terrace width. Under the chosen step-flow conditions, homoepitaxial growth of silicon on this initial surface leads to an increase of the step separation, as well as step edge roughening. A likely explanation is that step bunching occurs while there remain regions of ordered steps, with a smaller "local miscut". Whether steps remain of monolayer height or form bilayer high steps in these local staircases is not clear and requires further investigation.

Grazing incidence x-ray scattering proves to be a powerful tool to study the step topography at surfaces and interfaces; in particular for obtaining precisely the value and in-plane orientation of very small miscuts (~0.03°), and detailed statistical information on step distributions at surfaces.

[1] H.J. Gossmann, F.W. Sinden and L.C. Feldman, J.Appl. Phys. 67, 745 (1990).
[2] P.O. Hahn, M. Grunder, A. Schnegg and H. Jacob, Appl. Surf. Sci. 39, 436 (1989).
[3] P.H. Fuoss and I.K. Robinson, Nucl. Instr. and Methods, A222, 164 (1984).
[4] I.K. Robinson, Phys. Rev. B 33, 3830 (1986).
[5] G. Renaud, P.H. Fuoss, J. Bevk, B.S. Freer and P.O. Hahn, submitted to Phys. Rev. B.
[6] A. Mak, K.W. Evans-Lutterodt, K. Blum, D.Y. Noh, J.D. Brock, G.A. Held and R.J. Birgeneau, submitted to Phys. Rev. Lett. (1990).

Structure and Phase Transitions of Ge(111) and Si(111) Surfaces at High Temperatures

K.I. Blum[1], *D.Y. Noh*[1], *A. Mak*[1], *K.W. Evans-Lutterodt*[2], *J.D. Brock*[1,3], *G.A. Held*[4], *and R.J. Birgeneau*[1]

[1]Department of Physics, Massachusetts Institute of Technology, Cambridge, MA 02139, USA
[2]AT&T Bell Laboratories, Murray Hill, NJ 07974, USA
[3]School of Applied and Engineering Physics, Cornell University, Ithaca, NY 14853, USA
[4]IBM T.J. Watson Research Center, Yorktown Heights, NY 10598, USA

Abstract. Using glancing angle x-ray diffraction we have observed the long-range correlations of clean semiconductor surfaces with vacancies and steps. The Ge(111) surface undergoes a progressive disordering near 1070 K which we have found to be consistent with a model involving a proliferation of vacancies in the topmost bilayer. A miscut Si(111) surface is expected to be composed, at temperatures above the 7 x 7 reconstruction phase transition, of coherent steps which wander without crossing on the flat Si(111) surface. We have measured the structure factor of this surface and found that it indeed is well modeled by a rough surface with an algebraically decaying correlation function.

1. Introduction

Local surface probes have described the short-range structure of surface defects with great success [1]. Our work demonstrates the complementary information that x-ray provide on the long-range structure of surfaces with defects; this information is crucial for statistical mechanical models [2,3]. We have investigated the structure of two semiconductor surfaces: Ge(111) and vicinal Si(111). The first part of this paper will describe the disordering of the Ge(111) surface by the vacancies on the (111) terraces. The second part of this paper will describe the subtle disorder of the Si(111) surface containing a fixed concentration of steps.

2. Disordering of the Ge(111) Surface at High Temperatures

The Ge(111) surface was reported to undergo a phase transition at about 1050 K, which is 160 K below the bulk melting temperature [4]. These LEED results were interpreted on the basis of a model in which the outermost double layer of the Ge(111) surface consists of incommensurate crystalline islands surrounded by a disordered sea. Electron-energy-loss spectroscopy done on a Ge(111) single-crystal surface, suggested that the high-temperature disordered Ge(111) surface resembles an amorphous layer [5]. Our x-ray results show that vacancies produced in the (111) terraces account for these observations [2].

We have performed grazing incident x-ray scattering experiments on the IBM-MIT beam line X20 at the National Synchrotron Light Source at the Brookhaven National Laboratory. The resolution function was set by pairs of slits which were adjusted to fit experimental requirements. The half-width-at-half-maximum (HWHM) in the longitudinal direction was 2.2×10^{-3}Å^{-1} and in the transverse direction was 4.6×10^{-4}Å^{-1}. The Ge wafers were resistively heated inside a surface chamber that had a base pressure of 5×10^{-10} torr. The sample surface was prepared by ion bombardment (700eV, Ar^+) for 30 minutes at 970 K followed by one hour annealing at the same temperature. Auger-electron spectroscopy was done *in situ* to check surface cleanliness. The sample temperature was measured by two W5%Rh/W26%Ph thermocouples which were calibrated against an optical pyrometer.

Our study of the surface structure focused on the temperature dependent lineshapes of the (10) and (20) bulk-forbidden surface peaks. Representative (10) peaks at temperatures near 1050 K are shown in Fig. 1(a). There was no significant broadening or diffuse scattering in the tails throughout the temperature range up to near bulk melting. The peaks were always resolution limited indicating that no thermally generated steps with a mean spacing much less than 2000 Å^{-1} existed. This immediately ruled out surface roughening and lattice-gas melting models for a surface transition. The integrated intensities of the (10) and (20) surface peaks are shown in Fig. 1.(b). The intensities of these peaks dropped sharply up to about 1070 K and then exhibited an unexpected recovery at higher temperatures. We believe that the minimum is due to an interference between the scattering from successive bilayers. The similarity of the (10) and (20) peaks,

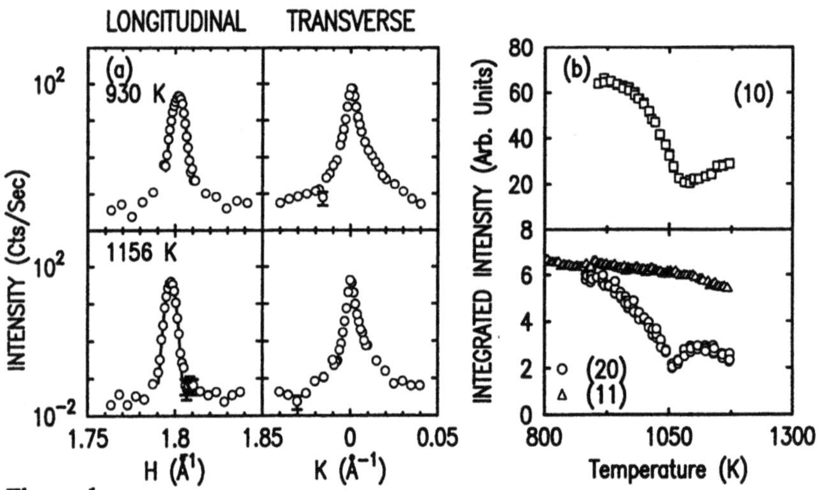

Figure 1

(a) Longitudinal(H) and transverse(K) scans through the (10) peak at 930 K and 1156 K. The central portion of longitudinal scans is fit to a Gaussian curve. (b) The integrated intensities of the (10), (20), and (11) peaks. The (11) peak shows the bulk Debye-Waller factor of Ge. The overall vertical scales are arbitrary.

the absence of any measurable diffuse scattering data, and the persistent long range order rule out models based on a surface melting transition near 1070 K.

We have proposed a model based on the effects of random vacancies on the structure factor of a bulk forbidden Bragg peak [2]. In our model we assume that evaporation of Ge atoms maintains a steady state with a constant concentration of missing surface atoms at a fixed temperature. At each temperature the bilayers are characterized by a vacancy fraction, $V_n(T)$ and a surface enhanced Debye-Waller (DW) factor, $M_n(T)$. The scattering amplitude from the nth bilayer $\alpha_n(\vec{q},T)$ can then be written as $\alpha_n(\vec{q},T) = [1-V_n(T)][1+e^{i\phi}]e^{-M_n(\vec{q})T}$, where ϕ is the phase shift between the layers comprising a bilayer. It is also the phase shift between successive bilayers. The intensity is given by the equation:

$$I = I_0 \left| \sum_{n=0}^{\infty} \alpha_{n+1}(q,T) e^{in\phi} e^{-n\varepsilon} \right|^2 ,$$

where I_0 is a normalization constant, and ε is the extinction length in units of the bilayer separation. We have assumed that the vacancies are uncorrelated (our data exclude correlation lengths exceeding about 20 Å) and we have used physically reasonable surface and bulk DW factors in our calculations. Fig. 2(a) shows calculated intensities using an exponentially decaying vacancy concentration, $V_n(T) = V_1 e^{-(n-1)/\xi}$ for various ξ's. Only when ξ is close to zero was the observed dip and recovery produced. This implies that most of the vacancies are confined to the topmost bilayer. Any other distribution of V_n with slower decay cannot produce the observed behavior of the intensities. The vacancy curves deduced from the (10) and the (20) intensity curves assuming all the vacancies are confined to the topmost bilayer are presented in Fig. 2(b).

One other possibility is the role of vacancies confined only to the first half bilayer. In this case the intensity is calculated using $\alpha_1 = [1-V_1^1(T)+e^{i\phi}]e^{-M_1T}$ and

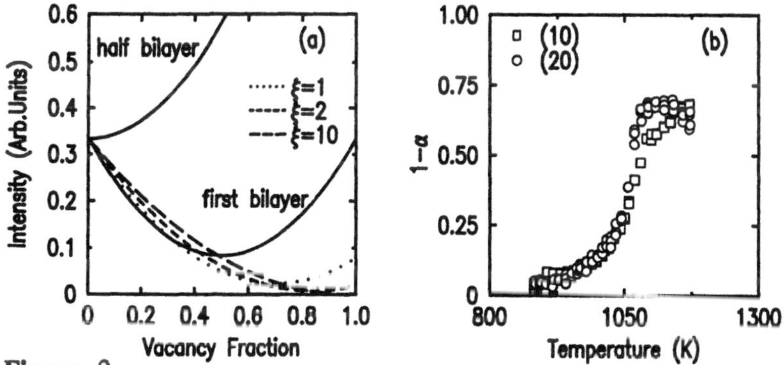

Figure 2
(a) Calculated intensities as functions of vacancy fraction. Vacancies confined to the first half bilayer give a monotonic increase in the intensity while those in the first bilayer give a dip and recovery structure. The three dashed lines are the results of calculation assuming $V_n = V_1 e^{-(n-1)/\xi}$. (b) The vacancy fraction calculated by the inversion of the equation in the text. All of the vacancies are assumed to be confined to the first bilayer.

$\alpha_n = [1+e^{i\phi}]e^{-M_1 T}$ for $n \neq 1$. As shown in Fig. 2(a) this produce a monotonic increase in the intensity. We also have calculated the effects of adatoms. The contribution of adatoms depends on the sites where adatoms sit. Adatoms on T_4 sites (on top of the atoms in the second bilayer) and T_1 sites (on top of the atoms in the first bilayer) only produces a monotonic increase in the intensity. Only adatoms on H_3 sites (on top of the atoms in the third bilayer) sites can generate the dip and recovery behavior similar to the one we observed. The minimum occurs when the coverage of adatoms is about half the monolayer. To obtain the recovery of the intensity the number of adatoms must increase progressively with increasing temperature. This is physically implausible.

We infer from our data that at high temperatures Ge(111) reaches a steady state with a large vacancy concentration in the surface layer. Although we are unable to probe directly the structure factor of uncorrelated point-like vacancies on the surface, we are able to infer the presence of vacancies from their characteristic effect on surface peaks.

3. Structure of Si(111) Surface above the 7x7 Reconstruction Transition.

Linear defects free to wander in two dimensions but not to cross produce a sharp feature in the structure factor which has power-law tails [6]. This effect has been observed for domain walls in the incommensurate stripe domain phase near the transition to the commensurate phase [7] Our observation of the same behavior for steps on Si(111) [3] will be described here.

A remarkable set of LEED [8] and STM [1] experiments demonstrated that vicinal Si(111) surfaces can exhibit a reconstruction driven faceting transition. In the high temperature uniform phase peaks due to an array of single bilayer steps were observed. Our own work is an attempt to probe quantitatively the long-range behavior of coherent surface structures, and to characterize the predicted roughness of the high temperature phase. The vacuum, thermal, and resolution conditions were all somewhat better for the silicon experiments than for the germanium experiments.

At temperatures above 1110 K we observed two step superlattice peaks bracketing the (10) position of the (111) surface. Fig. 3 shows high resolution longitudinal and transverse scans through the first order step peak at 1120 K. The simplest model leads us to believe that this peak results from the ordering of steps by the entropic repulsion. To test this idea quantitatively we fit the high temperature peak profile to a 2D power law singularity of the form

$$S(\vec{q}) = (q_H^2 + Bq_K^2)^{(-2+\eta_1)/2}$$

convolved with the instrumental resolution function. Fig. 3 also shows the optimal simultaneous fits to the longitudinal and transverse scans with η=0.6±0.1 and B = 15±7. It is evident that the power law form fits the data very well. The large value of η_1 is fully consistent with scattering from a rough surface whose height variance diverges logarithmically with distance. We note that as in the stripe domain commensurate-incommensurate problem the Pokrovsky-Talapov model leads to the prediction η_1=2/9, independent of temperature. This is in

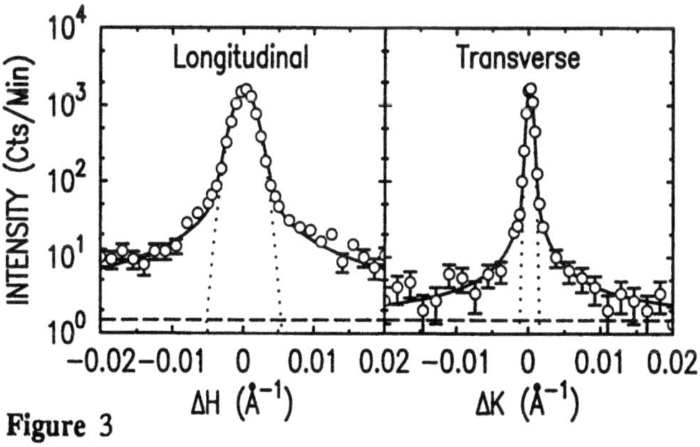

Figure 3
Longitudinal and transverse scans through the first order step surface peak. The dotted curves are the resolution. The solid curves are the result of fits to the power-law lineshape with η=0.6 and B=15.

clear disagreement with our results. More elaborate models of the interaction between the steps are needed to explain the large value of η_1.

The work at MIT was supported by the Joint Service Electronics Program under Contract No. DAAL-03-89-C-001.

References

1. E. D. Williams and N.C. Bartlett, Science Magazine **251**, 393 (1991).
2. A. Mak, K. W. Evans-Lutterodt, K. I. Blum, D. Y. Noh, J. D. Brock, G. A. Held, and R. J. Birgeneau, Phys. Rev. Lett. **66**, 2002 (1991).
3. D. Y. Noh, K. I. Blum, M. J. Ramstad, and R. J. Birgeneau, : (submitted for publication).
4. E. G. McRae, R. A. Malic, Phys. Rev. B **38**, 13163 (1988).
5. S. Modesti and A. Santoni, Solid State Commun., **72**, 315 (1989).
6. H. J. Schultz, Phys. Rev. B **22**, 5274 (1980).
7. S. G. J. Mochrie, A. R. Kortan, P. M. Horn, and R. J. Birgeneau, Z. Phys. **62**, 79 (1985).
8. R. J. Phaneuf and E. D. Williams, Phys. Rev. Lett. **58**, 2563 (1987).

Anomalous Scattering Applied to Co/Si(111) Interface Structure

N. Jedrecy[1,2], *A. Waldhauer*[1,3], *M. Sauvage-Simkin*[1,2], *R. Pinchaux*[1,4], *and V.H. Etgens*[1]

[1]LURE, CNRS-MENJS-CEA, Bat. 209d, F-91405 Orsay, France
[2]Laboratoire de Minéralogie-Cristallographie, 4 place Jussieu, F-75252 Paris Cedex 05, France
[3]Centre d'Etudes Nucléaires, SPAS, F-91991 Gif-sur-Yvette, France
[4]Université Pierre et Marie Curie, 4 place Jussieu, F-75252 Paris Cedex 05, France

Abstract. The anomalous scattering effect is used in a grazing incidence X-ray diffraction (GIXD) experiment to perform structural analysis of the Co/Si(111) 7x7 reconstructed interface. Data have been collected at different wavelengths, on both sides of the Co absorption K edge. The 7x7 ordering of the Co chemisorbed atoms is confirmed by variations in intensity of fractional order Bragg reflexions when changing X-ray energy.

The cobalt silicon interface offers the possibility of numerous fundamental and technological studies. Quasi perfect epitaxial $CoSi_2$ film may be formed by room temperature deposition of Co on top of a Si(111) substrate followed by simple annealing at 900°K ; a very sharp interface may be obtained. In the monolayer (ML) coverage range, a $CoSi_2$-like phase is already formed at room temperature, while higher coverages (up to 10 ML) lead to a mixture of $CoSi_2$-like and Co richer silicide phases (CoSi and Co_2Si) with an almost pure metallic Co film on top of the transition zone [1]. A detailed analysis of the very early stages of the Co film development should give a better understanding of the mechanisms of silicide formation. We report here on an X-ray diffraction study of a half-monolayer of cobalt deposited on the Si(111) 7x7 surface at 300°K. We assess that part of the chemisorbed Co atoms are long-range ordered with the same 7x7 periodicity than the bare Si substrate.

The Si(111) wafers were Shiraki-etched and cycle-annealed up to 1300°K in a UHV chamber (low 10^{-10} mbar) in order to obtain a sharp 7x7 LEED pattern. Co was deposited with a quartz oscillator calibrated electron gun evaporator, and the Co/Si ratio was controlled by Auger spectroscopy. For deposits in the ML range, we still observe the 7x7 LEED pattern blurred by diffused intensity.

Recently, SEXAFS measurements [2] have led to a chemisorption model where part of the Co atoms occupy six-fold interstitial sites (II6) within the top Si(111) bilayer (see fig. 1), the remaining part settling on top of the buried Co atoms adopting a three-fold coordination with the topmost Si atoms (T4 sites). An interesting feature of this model is that the chemisorption sites would not be the nucleation sites of $CoSi_2$. Moreover, the disappearance of the 7x7 streaks in the LEED pattern for a 1 ML deposit can be understood if one considers that the 7x7 Dimer-Adatom-Stacking fault (DAS) model [3] leaves only 20 H6 sites, and consequently 20 T4 sites per surface unit cell, whose saturation corresponds to a 0.8 ML Co deposit. As SEXAFS is a local order

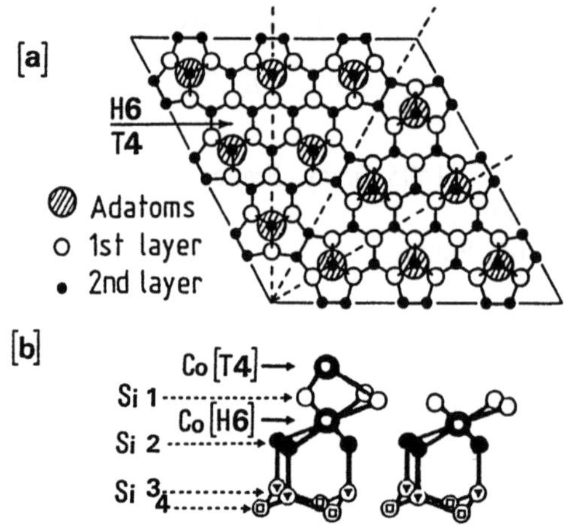

Fig. 1. [a]. Projection of the Si(111) 7x7 unit cell in the DAS model.
[b]. Side view along a <211> direction of the model (ref. 2) for chemisorption of Co in Si 7x7 H6 and T4 sites.

sensitive technique, it was important to control if the Si 7x7 structure was really governing the Co chemisorption.

A sample with a 0.5 ML Co deposit was transferred from the evaporation chamber to the coupled UHV diffractometer. Classically, for solving a reconstructed surface atomic structure, GIXD experiment consists in collecting as many as possible in-plane fractional order Bragg reflexions, in view of performing a refinement by comparison between experimental and model calculated structure factor amplitudes. In the case of large surface unit cells, many reflexions have to be measured, and the surface deterioration has to be taken into account. Another point is the difficulty to extract possible Co contribution to the diffracted intensities, since 0.5 ML corresponds only to 24 Co atoms compared to at least 90 Si atoms involved in the 7x7 reconstruction (excluding adatoms). Moreover, we may have to consider an unknown fraction of modified 7x7 surface areas with no ordering at all of the Co atoms. Measuring intensities of a given reflection at different wavelengths, at, below, and above the Co K edge (1.608 Å), and thus varying only the cobalt scattering factor, provides us direct evidence that Co atoms are involved in the 7x7 reconstruction with long-range ordering. Furthermore, an analysis based on the evolution of intensity with X-ray energy allows us to eliminate the problem of reconstructed areas without 7x7 ordered Co atoms. We may also determine some unknown terms in the structure factor by considerations of differences in intensity, making easier the final refinement.

We made some measurements at -24.4, 0, +19, +38, and +57 eV from the Co K edge. We used a symmetrical Si(111) channel-cut crystal as a monochromator ; the energy was controlled by recording the X-ray absorption from a pure Co film. Due to the vertical size and divergence of the X-ray source, our resolution in energy was 2.8 eV at Co K edge (7709 eV). We measured fluorescence from the sample during an energy scan of almost 150 eV to extract the energy correlated values of the dispersion correction terms f' and f" in the Co scattering factor. In-plane Fhk values were obtained by integration of the recorded intensity during sample rocking scans. Data were first

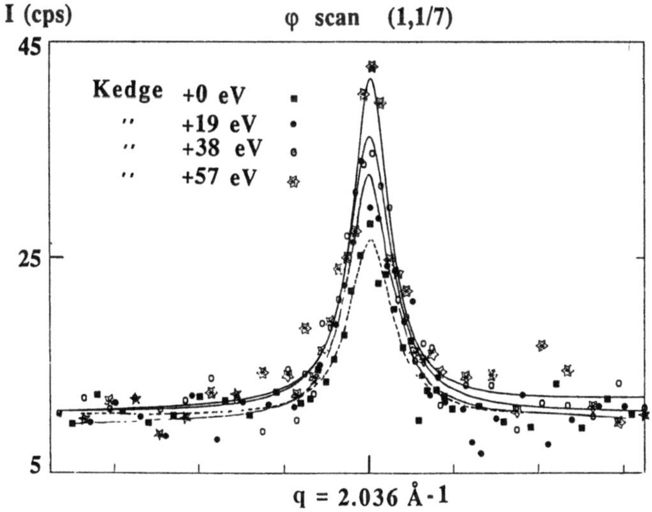

Fig. 2. Rocking scans about node (1, 1/7, l= 0.014) for 4 different X-ray energies at and above the Co absorption K edge. The scan performed at $E_{K_{Co}}$ -24.4 eV gave an intensity similar to the one at $E_{K_{Co}}$ +38 eV but is not shown in the graph for sake of clarity. The FWHM was of 0.00426 $Å^{-1}$.

collected at a constant grazing incidence angle from the 7x7 reconstructed Si(111) surface, then the sample was transferred and cleaned again before Co deposition to avoid possible contamination effects. The surface coherent domain size derived from the FWHM of the fractional order peaks decreased from 3000 Å to 1500 Å. The (1,1/7) reflexion clearly exhibited a variation in intensity when setting X-ray energy at the different values mentionned above (see fig. 2) indicating that some Co atoms are chemisorbed with 7x7 ordering.

We have no evidence of such clear dependence on energy for other reflexions, the variations being often of the same order of magnitude than the error bars on scan-derived integrated intensities. But, due to beam time shortage, this preliminary set of data is limited to 5 independent reflexions for which all the energies could not be recorded, and calculations show that each reflexion has a specific behaviour. These calculations were performed assuming the persistance of the 7x7 DAS scheme, with stacking fault islands in the first layer, dimers in the second one, but allowing the possibility of adatoms disappearance. Various combinations of H6 and T4 Co occupied sites were tried, but neither model was fully convincing. One point is that ratios between Fhk experimental values at a given energy cannot be explained if we consider a unique atomic configuration for 7x7 areas. Clearly, they differ from the Si DAS model or the Co-modified arrangements expected values. For such small deposits, a reconstructed surface area without ordered Co atoms contributing to the 7x7 intensities may easily hide the effect of modifying the Co scattering factor. Moreover, we carried out a second experiment who gave us smaller variations with X-ray energy change. Thus, it appears that ordering of Co chemisorbed atoms is highly sensitive to the experimental conditions.

We plan to work in the future with other phases in the same system corresponding to higher Co coverages and to use larger differences in X-ray energy to emphasize variations on both f' and f" values. Nevertheless, we have demonstrated the feasability of using anomalous scattering in surface diffraction experiments which may facilitate chemisorption structural studies. Indeed, a refinement process based on both ratios between F(h,k) amplitudes at different (h,k) positions and ratios between F(E) at different energy values is clearly more powerful than classical refinement. The collection of EXAFS data by recording fluorescence from the chemisorbed atoms in the unfavourable geometry of the GIXD experiment was shown to be possible. It enables an accurate determination of the dispersion terms in the scattering factor for a given chemical environment and may bring complementary information on the atom coordinates (in-plane with diffraction, out-of-plane with EXAFS).

References

[1] J. Derrien, Appl. Surf. Sc. 41/42, 241 (1989)

[2] G. Rossi, A Santaniello, P. De Padova, X. Jin, and D. Chandesris, Europhys. Lett. 11, 235 (1990)

[3] K. Takanayagi, Y. Tanishiro, S. Takahashi, and M. Takahashi, Surf. Sc. 161, 367 (1985)

X-Ray Reflectivity Studies of Au Surfaces

D.L. Abernathy[1], *D. Gibbs*[2], *G. Grübel*[2], *K.G. Huang*[2,4], *S.G.J. Mochrie*[1], *B.M. Ocko*[2], *A.R. Sandy*[1], *and D.M. Zehner*[3]

[1]Physics Department, Massachusetts Institute of Technology, Cambridge, MA 02139, USA
[2]Physics Department, Brookhaven National Laboratory, Upton, NY 11973, USA
[3]Solid State Division, Oak Ridge National Laboratory, Oak Ridge, TN 37831, USA
[4]Present address: Materials Science Division, Argonne National Laboratory, Argonne, IL 60439, USA

Abstract. We briefly review the results of x-ray scattering studies of the $Au(001)$ surface between 300 K and the bulk melting temperature $Tm = 1337\ K$. The specular reflectivities of the $Au(001)$, (110), and (111) surfaces obtained at 300 K are compared.

1. Introduction

During the last thirty years, significant progress has been made in our understanding of the room temperature structures of metal and semiconducting surfaces by conventional surface science techniques (such as LEED, ion scattering, STM, and others). Considerably less is known about the structure and phase behavior of surfaces at elevated temperatures, for example, near bulk melting temperatures. Interest in temperature-dependent studies of surface structure has recently been increasing rapidly, in particular, with the application of x-ray scattering techniques using synchrotron radiation.[1] Temperature-dependent studies provide insight into two-dimensional phase transformations. In addition, they are essential for an understanding of the global stability of crystal surfaces, and of the relation between such phenomena as roughening transitions, enhanced surface vibrations, and surface and bulk melting. During the last three years, we have investigated the structure and phase transformations of the low index faces of Au and Pt surfaces by x-ray glancing incidence and reflectivity techniques. We have by now completed studies of the $Au(001)$,[2] $Au(111)$,[3] $Pt(001)$,[4] and $Pt(111)$[5] surfaces in vacuum. The phases of the $Au(001)$ surface have also been investigated under electrochemical conditions.[6] In this paper, we summarize our results for $Au(001)$ in vacuum, emphasizing, in particular, the studies of the specular and non-specular reflectivities. We also compare the specular reflectivities of the $Au(001)$, (110), and (111) surfaces obtained at 300 K.

2. Au(001)

Experiments on clean Au surfaces, performed in an ultrahigh-vacuum environment, were carried out on beam lines X22 and X20, at the National Synchrotron Light Source. For large incidence angles (α), slits were used to define the illuminated sample area and all scattered x-rays were collected. Then the reflectivity from a given surface periodicity (τ_x, τ_y) is [2-3]

$$R = \left[\frac{4\pi r_0}{c^2 k \sin\alpha}\right]^2 |F(Q)|^2 e^{-2W(Q)} \left|\sum_{n=0}^{\infty} \rho_Q(n) e^{iQ_z cn/2}\right|^2, \qquad (1)$$

where $Q = k - k' = (\tau_x, \tau_y, Q_z)$ is the wave-vector transfer, r_0 is the Thomson radius, $F(Q)$ is the form factor, $W(Q)$ is the Debye-Waller factor, c is the appropriate lattice constant of Au and $\rho_Q(n)$ is the amplitude of the density wave in the nth layer. It is understood in the following that Q_z is along the surface normal direction. We will discuss specific forms for $\rho_Q(n)$ below. However, it is important to note that R is not sensitive to instrumental parameters, so that absolute measurements are possible.

We have identified three distinct structural phases which are exhibited by the $Au(001)$ surface between $T = 300\ K$ and the bulk melting transition at $T = 1337\ K$. They are (1) disordered ($1170 < T < 1337\ K$), (2) corrugated hexagonal ($970 < T < 1170\ K$), and (3) rotated, corrugated hexagonal ($300 < T < 970\ K$). The symmetries of the diffraction patterns for these phases are shown in Figure 1, where the data obtained in the glancing-incidence geometry are summarized. In the figure, each symbol represents a rod of scattering extending normal to the (001) planes. The rod scattering through the origin of reciprocal space is the specular reflectivity, while the off-axis rod scattering is the nonspecular reflectivity.[2,4]

The diffraction pattern observed in the high-temperature phase between $T = 1170$ and $1337\ K$ is shown in Figure 1(a). Solid squares are indexed in bulk-cubic reciprocal-lattice units $(H, K, L)_c$ and represent the reflectivity along directions normal to the surface joining all the bulk Bragg reflections. In the literature, these rods are referred to as crystal-truncation rods.[7] Thus, the area enclosed by the solid squares corresponds to the substrate unit cell. At elevated temperatures the in-plane diffraction pattern has the (1×1) symmetry of the bulk, consistent with an unreconstructed or disordered surface layer. More importantly, the Q_z-dependence of the reflectivity is consistent with the existence of a thin disordered film (1–2 layers thick), and not with an ideally terminated crystal. Schematic real space side and top views of this phase of the $Au(001)$ surface are shown in Figure 2(a).

Below $T = 1170\ K$ there is a reversible transition to an incommensurate, two-dimensional structure of hexagonal symmetry. Hexagonal reciprocal-lattice vectors $(1, 0)_h$ and $(0, 1)_h$ are indicated by open triangles in Figure 1(b). An important feature of this structure is an incommensurate corrugation along the cubic [110] direction with a wave vector (δ) just equal to the difference between overlayer $(1, 0)_h$ and substrate $(1, 1, 0)_c$ wave vectors. In Au reciprocal-lattice

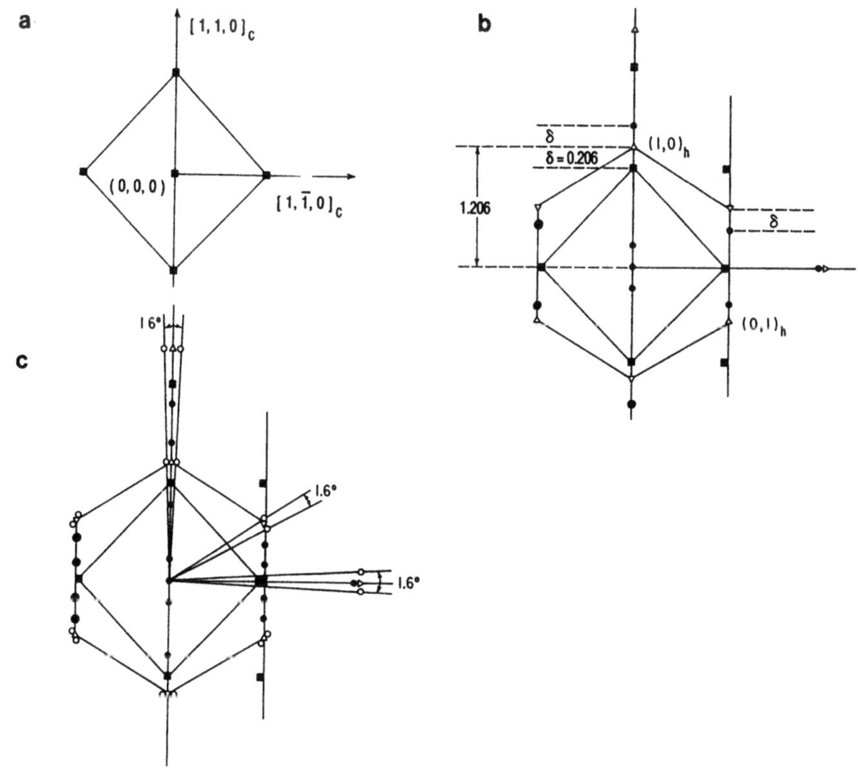

Figure 1. In-plane diffraction pattern $(H,\ K,\ 0.07)_c$ for (001) surface of *Au*. (a) Disordered phase, (b) corrugated-hexagonal phase, and (c) rotated, corrugated-hexagonal phase. Each point represents a rod of scattering extending normal to the surface. Bulk [110]c and $[1\bar{1}0]$c directions are indicated. Peaks corresponding to domains rotated by 90° were observed, but are not shown. All the data represented in Figure 1 were obtained at fixed $L = 0.07$ in bulk reciprocal-lattice units.

units, $\delta = (0.206 \pm 0.001)\sqrt{2}c^\star (c^\star = 2\pi/c$, where $c = 4.08$ Å at 300 K). This gives rise to satellites (solid circles) along directions parallel to the [110] direction, about each hexagonal position. These satellites arise from modulations of the surface layer by the underlying square lattice, and vice-versa. Real space schematic views of this phase, which we call the corrugated-hexagonal phase, are shown in Figure 2(b). The incommensurability in the $[1\bar{1}0]$ direction is $(0.043 \pm 0.001)\sqrt{2}c^\star$, but no corresponding satellite peaks were observed.

Below $T = 970\ K$, additional rods appear around each of the hexagonal rods, as shown by the open circles in Figure 1(c). These occur at a fixed angle equal to ±0.81°, and reveal the existence of rotated domains (in coexistence with the aligned domains).[2,4] In addition, higher harmonics are visible along the [110] direction (solids circles, Figure 1(c)). The transition to the rotated, corrugated-hexagonal phase is reversible. Figure 2(c) shows a schematic real-space view of the rotated, corrugated-hexagonal phase.

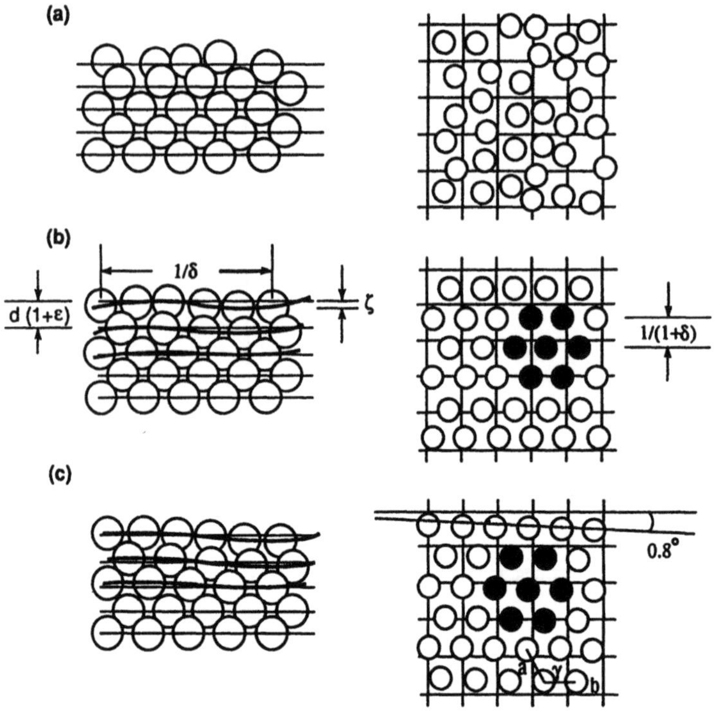

Figure 2. Schematic real space side (left) and top views (right) of the observed phases of the $Au(001)$ surface. (a) Disordered phase, (b) corrugated-hexagonal phase, (c) rotated, corrugated-hexagonal. On the left is depicted the kind of information learned in reflectivity measurements, while on the right is the kind obtained at glancing incidence. ζ is corrugation amplitude, δ is the incommensurability, d is the interlayer spacing of the ideally terminated bulk, and ϵ is the fractional change in interlayer spacing upon surface relaxation.

Figure 3 displays the measured absolute reflectivity as a function of $L = Q_z c/2\pi$ at a temperature of 1100 K, for several in-plane wave vectors. Because the data extend over a large range to $Q_z \cong 5\mathring{A}^{-1}$, they are particularly sensitive to atomic coordinates perpendicular to the surface, and allow detailed modeling of the surface structure. Each point represents the integrated intensity obtained in a transverse scan. The specular reflectivity is plotted in Figure 3, curve a. As has been discussed previously,[2,4] the line-shape asymmetry about the Bragg peak is a clear signature of a top-layer expansion [$d_{01} = 1.19 \times (c/2)$]. Figure 3, curves e and f, show the nonspecular reflectivity at the principle hexagonal wave vectors, $(1,\ 0)_h$ and $(0,\ 1)_h$. Qualitatively, the Q_z dependence follows the $(\sin\alpha)^{-2}$ behavior expected from Eq.(1) for a hexagonal monolayer. The small deviations are related to the surface corrugation and small displacements parallel to $(1\bar{1}0)$.[3,4] Figure 3, curve b, shows the reflectivity at the modulation wave vector $(\delta,\ \delta,\ L)$. This profile shows a significant increase from $L = 1$ to 2 and a line-shape asymmetry about $L = 2$, reminiscent of the specular reflectivity. The simplest explanation of the similarity of the Q_z-dependence

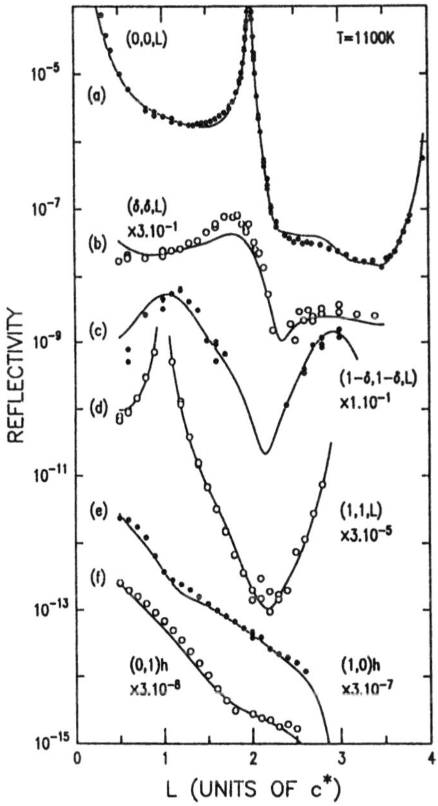

Figure 3. Absolute reflectivity of $Au(001)$ at 1100 K for various different in-plane wave-vector transfer (τ_x, τ_y): curve a (0, 0), specular reflectivity; curve b, (δ, δ); curve c, ($1-\delta$, $1-\delta$); curve d, (1, 1); curve e, $(0,\ 1)_h$, and curve f, $(1,\ 0)_h$.

of the reflectivity at an incommensurate in-plane wave vector (δ, δ) to the specular reflectivity is that several layers participate in the reconstruction; that is, are modulated. The absence of significant increase near $L = 0$ of the non-specular reflectivity along (δ, δ, L) further suggests that the modulation is a surface corrugation with atomic displacements normal to the surface. A schematic view of the inferred multilayer buckling is shown in Figures 2(b)and 2(c), Similar remarks apply also to the non-specular reflectivities along (1, 1, L) and ($1-\delta$, $1-\delta$, L) shown in Figures 3(c) and 3(d), respectively.

We have also fitted the reflectivity using Eq.(1) assuming a locally commensurate (5×1) structure. While this motif is not repeated periodically, nevertheless, such a simple model for $\rho_Q(n)$ provides an excellent description of the data (solid lines in Figure 3). The parameters used were a layer-dependent sinusoidal corrugation amplitude for each layer ($\zeta_0 - \zeta_5$), the layer spacings (d_{01}, d_{12}), and a layer-dependent vibrational amplitude normal to the surface ($\sigma_0 - \sigma_3$). A detailed expression for $\rho_Q(n)$, including these parameters, is given in reference [2]. Allowing parameters for additional layers to vary did not improve the

Table 1: The rms surface-normal vibrational amplitude for the nth layer (σ_n) vs temperature. The surface layer corresponds to $n = 0$.

Temperature (K)	σ_0 (Å)	σ_1 (Å)	σ_2 (Å)	σ_3 (Å)	σ_4 (Å)	σ_5 (Å)	Bulk (Å)
300	0.19 ± 0.02	0.17	0.09	0.09	0.09	0.09	0.09
1100	0.32	0.26	0.22	0.19	0.17	0.17	0.17
1200	0.44 ± 0.04	0.34	0.29	0.24	0.21	0.19	0.18

quality of the fits. Fitting all of the profiles of Figure 3 together confirms the expectations outlined above and describes in detail the data of Figure 3, curves c and d. We find that the dependence of ζ_n on the distance from the surface (Z_n) is well described by $\zeta_n = \zeta_0 \exp(-Z_n/\lambda)$ with $\zeta_0 = 0.28 \pm 0.02\mathring{A}$. A value of $0.56\mathring{A}$ peak-to-peak is in reasonable agreement with He atom scattering results,[8] and $\lambda = 3.3 \pm 0.6\mathring{A}$. Simple elasticity theory leads one to expect that λ is of the order of the modulation wavelength divided by 2π, i.e., $2.29\mathring{A}$. The surface-normal vibrational amplitudes approach the bulk value for large n but are enhanced near the surface (Table 1). Only the topmost interlayer spacings d_{01} is measurably different from the bulk.

At $T = 1170\ K$, all of the scattering at the in-plane hexagonal wave vectors, and of their satellites, abruptly disappears. In addition, the reflectivity near the (110)c-position decreases by nearly a factor of 100 as the temperature is increased above 1170 K. In order to explore the hexagonal disorder transition at high temperatures, we performed measurements of the specular and non-specular reflectivity near 1170 K. Examples are shown in Figure 4 at 1200 K and, for comparison, at 1100 K in the corrugated-hexagonal phase. Evidently, there is a dramatic change in going through the transition. In contrast, there is no change in the transverse line shape or position.[2,4] One may characterize the specular reflectivity as corresponding to an unreconstructed surface with enhanced surface-normal vibrational amplitudes. However, it is impossible on this basis alone to understand the decrease of the (1, 1, L) reflectivity at small L from 1100 to 1200 K. On the other hand, the reflectivity from a gradual interface is invariably less than that from a sharp interface. Thus these data imply that at high temperatures the in-plane (1, 1, 0) cubic density wave decays as the surface is approached from the bulk. This, in fact, is the behavior expected if the surface undergoes a melting transition.[9,10] To model the data of Figure 4, we have chosen the following form for $\rho_G(n)$ [$G = (110)$ cubic] in the disordered phase.

$$\rho_G(n)/\rho_G(\infty) = 1 - 1/[1 + e^{\kappa \ell} sinh(\kappa n + S)], \tag{2}$$

with $e^{\kappa\ell}[\ 1/(1 - e^{-\kappa\ell}) - 1]$. Other parameterizations are also possible.[2] According to Eq.(2), the density-wave profile is described by two parameters: the width of the interface κ^{-1} and the thickness of the disordered, surface layer (2).

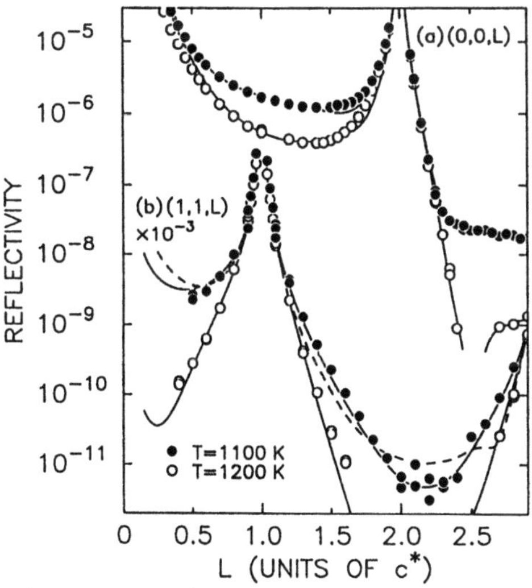

Figure 4. (a) Specular and, (b), nonspecular reflectivity at 1100 K (closed circles) in the corrugated-hexagonal phase and at 1200 K in the disordered phase (open circles). Solid lines are fits to the reflectivity discussed in the text.

The solid lines of Figure 4 are the results of fits with a model which allows a surface-normal vibrational amplitude for each layer near the surface and which characterizes $\rho_G(n)$ via Eq.(2); the density of the surface layer was fixed equal to that of the bulk planes. The resultant values of κ^{-1} and ℓ are $2.9 \pm 0.4Å$ and $3.2 \pm 1.2Å$, respectively, and indicate that the surface may be described as one or two disordered layers at these temperatures (see Figure 2(a)). The best-fit value of d_{01} corresponds to an 8.5% expansion. The vibrational amplitudes, determined by the specular reflectivity, are given in Table 1. Since the surface-normal vibrational amplitudes are less than the (001)-layer spacing, the layer order is still well established right to the surface. Surface disordering transitions have previously been observed on $Pb(110)$.[11]

3. Au(111), Au(001), and Au(110).

Finally, in Figure 5 we show the specular reflectivities measured versus momentum transfer for the $Au(111)$, (001), and (110) surfaces obtained at 300 K. The sharp peaks in each profile correspond to the various Bragg reflections: (111) and (222) for the $Au(111)$ surface, (002) for the $Au(001)$ surface, and (220) for the $Au(110)$ surface. The solid lines are fits of the specular reflectivities to Eq.(1). (The undistorted layer spacing c_n is $a/2$, $a/\sqrt{3}$, or $a/\sqrt{2}$ for the (001), (111), and (110) surfaces, respectively). In each case, these surfaces exhibit close-packed reconstructions within the surface plane, which lead to modified values of the average surface density (compared to their respective bulk planes) and to the possibility of surface buckling. In addition, there may exist a relaxation of the top and subsequent layers along the surface normal

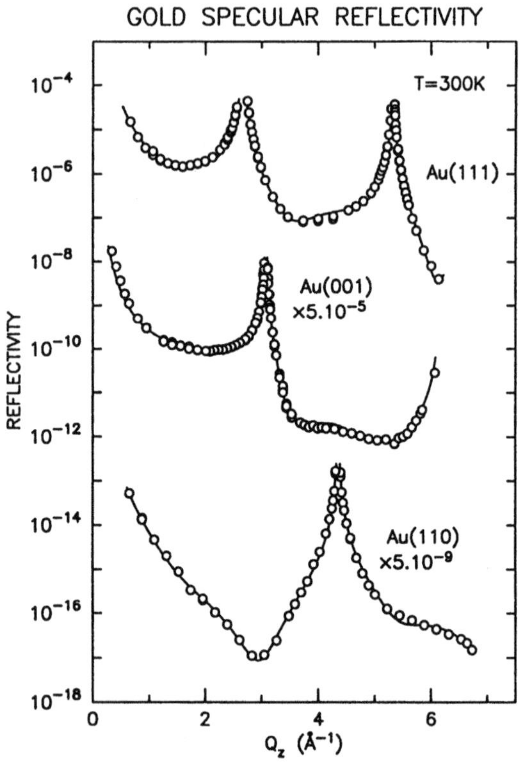

Figure 5. Specular reflectivities of the $Au(111)$, (001), and (110) surfaces versus momentum transfer Q_z. These data were obtained at 300 K.

direction, leading to modified values of the interplanar spacings (d_{01}, d_{02}, etc.) In the discussion of Figure 3 above, we noted that the asymmetry of the reflectivity above and below the (002) reflection for the $Au(001)$ surface arises mainly from a 20% outward expansion of the top layer along the surface normal direction. To obtain the fits shown in Figures 3 and 5 for $Au(001)$, it has also been necessary to include a 25% increase of the top layer density (consistent with a hexagonal overlayer) and a 19% buckling amplitude.[2] In contrast, the reflectivity obtained from the $Au(110)$ surface is asymmetric with the larger intensity appearing at higher momentum transfer, consistent with an inward contraction of the top layer spacing. Fits of Eq.(1) to this profile give a top layer density of 0.50 ± 0.03 and a contraction of $17.6 \pm 3.6\%$, relative to the bulk (110)-planes. A top layer density of 0.5 is consistent with a (2×1) reconstruction for the $Au(110)$ surface. Measurements of the in-plane diffraction pattern for this surface have confirmed the (2×1) reconstruction. The large contraction is consistent with the results of LEED, ion scattering, and recent x-ray diffraction experiments.[12] For the $Au(111)$ surface, the best fits resulted in a top layer density increased by 4% and a small outward expansion of $3 \pm 0.4\%$ of the top layer. These results are consistent with the chevron reconstruction discussed in reference [3].

In summary, we have briefly described our results of x-ray diffraction experiments on the $Au(001)$ surface between 300 K and the bulk melting temperature of $Tm = 1337\ K$. We have identified three distinct structural phases; (1) disordered ($1170\ K < T < 1337\ K$), (2) incommensurate, corrugated-hexagonal ($970\ K < T < 1170\ K$), and (3) incommensurate, rotated, corrugated-hexagonal ($300\ K < T < 970\ K$). We have also compared the specular reflectivities at 300 K of the $Au(001)$, (110), and (111) surfaces. The qualitative differences observed in the profiles are consistent with the three different reconstructions of these surfaces.

4. Acknowledgements

We would particularly like to thank G. Ownby for technical assistance. Work at MIT is supported by the NSF (Grants No. DMR-8806591 and No. DMR-8719217), work at BNL by the U.S. DOE (Contract No. DE-AC02-76CH00016), and at ORNL by the Division of Materials Sciences, U.S. DOE under (Contract No. DE-AC05-84OR21400) with Martin-Marietta Energy Systems Inc. X20 is supported by the NSF Materials Research Laboratory Program MIT (Grant No. DMR-8719217) and by IBM.

5. References

[1] See papers in this proceedings.

[2] S.G.J. Mochrie, D.M. Zehner, B.M. Ocko, and D. Gibbs, Phys. Rev. Lett. **64**, 2925 (1990); D. Gibbs, B.M. Ocko, D.M. Zehner, and S.G.J. Mochrie, Phys. Rev. B **42**, 7330 (1990); B.M. Ocko, D. Gibbs, K.G. Huang, D.M. Zehner, and S.G.J. Mochrie, Phys. Rev. B (in press)

[3] K.G. Huang, D. Gibbs, D.M. Zehner, A.R. Sandy, and S.G.J. Mochrie, Phys. Rev. Lett. **65**, 3313 (1990); A.R. Sandy, S.G.J. Mochrie, D.M. Zehner, K.G. Huang, and D. Gibbs, Phys. Rev. B **43**, 4667 (1991).

[4] D. Gibbs, G. Grübel, D.M. Zehner, D.L. Abernathy, and S.G.J. Mochrie, Phys. Rev. Lett. (submitted); D.L. Abernathy, S.G.J. Mochrie, D.M. Zehner, G. Grübel, and D. Gibbs, Phys. Rev. B (submitted); G. Grübel, D. Gibbs, D.M. Zehner, D.L. Abernathy, and S.G.J. Mochrie, (unpublished).

[5] A.R. Sandy, S.G.J. Mochrie, D.M. Zehner, G. Grübel, K.G. Huang, and D. Gibbs, (unpublished).

[6] B.M. Ocko, J. Wang, A. Davenport, H. Isaacs, Phys. Rev. Lett. **65**, 1466 (1990).

[7] I.K. Robinson, Phys. Rev. B **33**, 3830 (1986); S.R. Andrews and R.A. Cowley, J. Phys. C **18**, 6427 (1985); A.M. Afanaslev, P.A. Aleksandrov, S.S. Fanchenko, V.A. Chaplanov, and S.S. Yakimov, Acta Cryst., Sect. A **42**, 116 (1986).

[8] K.H. Rieder, T. Engel, R.H. Swendsen, and M. Manninen, Surf. Sci. **127**, 223 (1983).

[9] R. Lipowsky, V. Breuer, K.C. Prince, and H.P. Bonzel, Phys. Rev. Lett. 62, 913 (1989).
[10] H. Lowen, T. Beier, H. Wagner, Euro-Phys. Lett. 9, 701 (1989).
[11] J.W.M. Frenken, and J.F. Vanderveen, Phys. Rev. Lett. 54, 134 (1985); J.W.M. Frenken, P.M.J. Maree, and J.F. Vanderveen, Phys. Rev. B 34, 7506 (1986); B. Pluis, T.N. Taylor, D. Frankel, and J.F. Vanderveen, Phys. Rev. B 40, 1353 (1989)
[12] W. Moritz and D. Wolf, Surf. Sci. 88, L29 (1978); S.H. Overbury, W. Heiland, D.M. Zehner, S. Datz, and R.S. Thoe, Surf. Sci. 109, 239 (1981); J. Moller, H. Neihus, and W.Heiland, Surf. Sci. 166, L111 (1986); M. Copel and T. Gustafsso, Phys. Rev. Lett. 57, 723 (1986); E. Vlieg, I.K. Robinson, and K. Kerne, Surf. Sci. 233, 248 (1990)

Crystal Truncation Rod as a Convolution of Three-Dimensional Bravais Lattice with X-Ray Reflectivity

Hoydoo You

Argonne National Laboratory, Material Science Division,
9700 South Cass Ave., Argonne, IL 60439, USA

Abstract An expression for crystal truncation rod intensity is derived from a convolution of three dimensional Bravais lattice with x-ray reflectivity. For mosaic blocks with no miscut surface the expression recovers Robinson's result. For the mosaic blocks with miscuts, the expression is reduced to Andrews and Cowley's form.

1. Introduction

Andrews and Cowley[1] pointed out that x-ray scattering from a single crystal surface should exist near a Bragg reflection. Independently, Robinson[2] has shown that the scattering intensity from a single crystal surface is equal to a half infinite sum of point scatterers, known as the crystal truncation rod (CTR). While the former expression was derived for small $\mathbf{Q}-\mathbf{G}$ near Bragg reflections but far enough from the Darwin width, the latter approach is more general and has been widely used for surface x-ray scattering studies.[3] On the other hand, x-ray reflectivity based on the Fresnel approach of classical optics has been extensively used[4] to study various interfaces. Although it is widely recognized[5,6] that CTR should be derivable by convolution of the reflectivity and the Bragg reflections, it has not been written down so far. In this short article, therefore, we write down the equation in the form of convolution and will show that our expression becomes Robinson's form or Andrews and Cowley's form depending on the physical situation.

2. Review of X-ray Reflectivity

In this section, a brief review of the well known results for x-ray reflectivity is given for the later discussions. The amplitude of specularly reflected x-rays from an ideally sharp surface or interface for polarization perpendicular to the scattering plane is

$$R(Q) \;=\; \frac{Q_z - Q'_z}{Q_z + Q'_z} \approx \left(\frac{Q_c}{2Q_z}\right)^2 \tag{1}$$

where $Q'_z = \sqrt{Q_z^2 - Q_c^2}$ with $Q_c = 0.0289\sqrt{n}$ and n is the electron density. When the kinematic scattering limit is considered, the reflectivity can be obtained from the Fourier transformation of a step function, $\Theta[g(\mathbf{r})]$, where $g(\mathbf{r}) = 0$ defines the surface:

$$R(Q) = A(Q)\, n \int e^{i\mathbf{Q}\cdot\mathbf{r}}\Theta[g(\mathbf{r})]d\mathbf{r}\,. \tag{2}$$

This expression is essentially the surface integral given by Sinha et al..[5] For an ideally sharp interface the reflected amplitude becomes by using Dirac's identity,

$$R(Q) = A(Q)\, n \lim_{\varepsilon\to 0} \frac{1}{iQ+\varepsilon} = A(Q)\, n \left[\pi\delta(Q) - i\mathcal{P}\frac{1}{Q}\right]. \tag{3}$$

The delta function is analogous to the usual Bragg reflection for an imperfect crystal, and the second term, Cauchy principal value of $1/Q$, is the surface term. Since we are interested in CTR away from Bragg reflections the delta function will not be used in the later discussion. Equations (1) and (3) become similar at a large Q since the reflection becomes weaker and satisfy the kinematic scattering limit. The prefactor, $A(Q)$, results from refraction of x-rays and can easily be determined from eqs (1) and (3).

3. Crystal Truncation Rod

For a single crystal case, it is convenient to start with the density of an infinite single crystal in the form of a Bloch wave expansion.[7] Then the electron density of a half infinite single crystal can be expressed as a product of a step function and the sum of Block waves,

$$n(\mathbf{r}) = \Theta[g(\mathbf{r})] \sum_{\mathbf{G}} F(\mathbf{G})\, e^{i\mathbf{G}\cdot\mathbf{r}} \tag{4}$$

where $F(\mathbf{G})$ is a structure factor. Then the scattering amplitude in the kinematic limit

$$R(\mathbf{Q}) = A(\mathbf{Q}) \sum_{\mathbf{G}} F(\mathbf{G}) \int e^{-i(\mathbf{Q}-\mathbf{G})\cdot\mathbf{r}}\, \Theta[g(\mathbf{r})]\, d\mathbf{r} \tag{5}$$

$$= A(\mathbf{Q}) \sum_{\mathbf{G}} \frac{F(\mathbf{G})}{\mathbf{Q}_\perp - \mathbf{G}_\perp} \int_S e^{-i(\mathbf{Q}-\mathbf{G})\cdot\mathbf{r}}\, d\mathbf{S}_\perp\, . \tag{6}$$

Equation (5) is analogous to eq (2) and eq (6) is from Sinha et al.'s surface integral. Now we can easily see by comparing eqs (2) and (5) that CTR is amplitude sum of the reflections from the step density $F(\mathbf{G})$ and the momentum transfer $(\mathbf{Q}-\mathbf{G})_\perp$. Now the factor $A(Q)$ can be found by substituting $(\mathbf{Q}-\mathbf{G})_\perp$ and $(\mathbf{Q}'-\mathbf{G})_\perp$ for Q_z and Q'_z into eq (1) and compare it with eq (6) for a sharp interface. $\mathbf{Q}'_\perp$ is defined in the same manner as $\mathbf{Q}'_\perp = \sqrt{Q_\perp^2 - Q_G^2}$ with $Q_G = 0.0289\sqrt{F(\mathbf{G})}$. Finally, the CTR intensity becomes

$$\begin{aligned} I(\mathbf{Q}) &= |R(\mathbf{Q})| \\ &= A^2(\mathbf{Q}) \sum_{\mathbf{G}} \frac{|F(\mathbf{G})|^2}{(\mathbf{Q}_\perp - \mathbf{G}_\perp)^2} \int_S \int_{S'} e^{-i(\mathbf{Q}-\mathbf{G})\cdot(\mathbf{r}-\mathbf{r}')}\, d\mathbf{S}_\perp d\mathbf{S}'_\perp \\ &+ A^2(\mathbf{Q}) \sum_{\mathbf{G}\neq\mathbf{G}'} \frac{F(\mathbf{G})F(\mathbf{G}')}{(\mathbf{Q}_\perp - \mathbf{G}_\perp)(\mathbf{Q}_\perp - \mathbf{G}'_\perp)} \int_S \int_{S'} e^{-i(\mathbf{Q}-\mathbf{G})\cdot\mathbf{r}} e^{-i(\mathbf{Q}-\mathbf{G}')\cdot\mathbf{r}'}\, d\mathbf{S}_\perp d\mathbf{S}'_\perp\, . \end{aligned} \tag{7}$$

By examining these two terms, we can see that the second term is from the interferences among overlapped CTR's from several Bragg reflections. When the surface is miscut from a crystallographic plane, the second term can be ignored and the scattering intensity is reduced to Andrews and Cowley's form. When the crystal surface is cut parallel to a well defined crystallographic surface, all the interference terms should be necessarily included and the equation becomes Robinson's form.

4. Perfectly Cut Surface Case

In this section, we will consider a surface perfectly cut to a crystallographic plane. When the first plane of unit cells is at $z = 0$, the most physically reasonable position of cut plane is at $z = c/2$, a half unit cell away from the origin. From eq (6) the intensity becomes

$$
\begin{aligned}
I(\mathbf{Q}) &= \left| A(\mathbf{Q}) \sum_{\mathbf{G}} \frac{F(\mathbf{G}) e^{-i(\mathbf{Q}_\perp - \mathbf{G}_\perp)c/2)}}{\mathbf{Q}_\perp - \mathbf{G}_\perp} \right|^2 \\
&= \left| A_{H,K}(\ell) \sum_{L} \frac{F_{H,K}(L) e^{i\ell\pi}}{\ell - L} \right|^2 && (8) \\
&= \left| -A_{H,K}(\ell)\, Res \left[\frac{\pi F_{H,K}(z) csc(\pi z)}{\ell - z} \right] \right|^2 && (9) \\
&= \left| A(\ell) \pi F_{H,K}(\ell) csc(\pi \ell) \right|^2 . && (10)
\end{aligned}
$$

Equation (10) is now identical to CTR equation given in the reference.[3] The position of the cut surface is very important. If we chose the cut surface at the origin, the final function would have been $cot(\pi\ell)$ instead of $csc(\pi\ell)$. If the cut surface was at an arbitrary position, the summation may not have a closed form.

5. Conclusion

A few remarks need to be made. First, the step function Θ was used to derive the equation with an assumption of an ideally sharp interface. However, the step function can be replaced with another function, whose value is 0 at $-\infty$ and 1 at ∞, for a smooth interface or a reconstructed interface. Therefore the equation shown in this paper in principle can be generalized. Second, it is important to separately consider the two terms in eq (7). The second term (interference term) can be small or vanishing not because the mosaic surfaces are rough but because the mosaic surfaces are miscut. It is important to consider the miscut mosaic blocks through an incoherent average in the data analysis. Third, most importantly, our expression is much easier for the case with a large unit cell composed of several atoms. For the large unit cell case, the cut surface can be inside the unit cell itself. Because CTR is extremely sensitive function of the cut surface, it has recently been used for the top layer determination of $YBa_2Cu_3O_{7-\delta}$ single crystal surface.[8] Finally, it should be noted that our expression is strictly based on the assumption that the crystal of our interest has a true 3-dimensional long range order so that the electron density can be expressed in a form of Bloch wave expansion. Therefore, the definition of CTR used in this article breaks down for the material with a partial or quasi- long range order such as multilayers.

Acknowledgements

We thank I.K. Robinson, J.D. Axe, S.G.J. Mochrie, S.K. Sinha, D. Gibbs, and S.C. Moss for the useful discussions. We also thank R.P. Chiarello for reading of the manuscript. This work was supported by DOE under contract W-31-109-ENG-38.

References

1. S.R. Andrews and R.A. Cowley, J. Phys. C, **18** 6427 (1985).
2. I.K. Robinson, Phys. Rev. **B33**, 3830 (1986).
3. D.G. Gibbs, B.M. Ocko, D.M. Zehner, and S.G.J. Mochrie, Phys. Rev. **B38** 7303 (1988).

4. For example, I.M. Tidswll, B.M. Ocko, P.S. Pershan, S.R. Wassermann, G.M. Whitesides, and J.D. Axe, Phys. Rev. **B41**, 1111 (1990).

5. S.K. Sinha, E.B. Siroda, S. Garoff, and H.B. Stanley, Phys. Rev. **B 38** 2297 (1988).

6. Private communication.

7. B.W. Batterman and H. Cole, Rev. of Mod. Phys., **36**, 681 (1964).

8. To be published.

Extended X-Ray Reflectivity Analysis of Si(111)7×7

I.K. Robinson and E. Vlieg

AT&T Bell Laboratories, Murray Hill, NJ 07974, USA

A number of new results are emerging from the various techniques for measurement of surfaces using X-ray and neutron scattering methods, as witnessed by the program of this conference. With many names (and thankfully few acronyms) currently in use for these methods, it can be helpful to summarise them according to the regions of reciprocal space they refer to, as we attempt to do in Fig 1. This forms a useful guide to evaluating the level of information attained in each experiment, with regard to resolution of detail and its direction. The principal distinction is between the reflectivity ($Q_\parallel \approx 0$) and diffraction ($Q_\parallel >> 0$) regions; further subdivisions apply when $Q_\perp \approx 0$ or not, as shown. A finer distinction could have been made between the analysis of crystal truncation rods ("integer order" diffraction) and superstructure reflections (fractional orders), but we chose not to show this.

In this paper we concentrate on the interpretation of reflectivity data. The conventional view of X-ray reflectivity is understood to be represented by the Fresnel law, and the deviations from it which may have structural interpretations [1]. The *non-specular* reflectivity region is the surface equivalent of small angle scattering. Here the collective modes of fluid surfaces, in the form of capillary waves, are detected [2]. In the reflectivity region far from the origin where the crystalline Bragg peaks start to appear, the simple dielectric boundary description of the Fresnel law no longer applies. In order to separate its identity we refer to this as the *extended reflectivity* region.

In the extended reflectivity region the refraction effects near $Q_\perp = 0$ are unimportant, and the kinematic approach analogous to the analysis of crystal

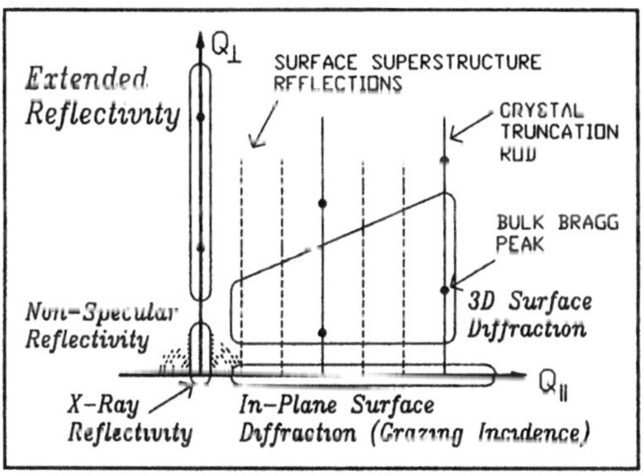

Figure 1. Schematic map of surface scattering methods in reciprocal space.

Springer Proceedings in Physics, Vol. 61 **Surface X-Ray and Neutron Scattering**
Eds.: H. Zabel and I.K. Robinson

truncation rods (CTRs) [3] is appropriate, except that all lateral order within the layers of the sample is invisible. The scattering amplitude can be represented as a one-dimensional sum over the layers of the crystal from the top (z = 0) to z = ∞. Quite generally we can represent each layer by an index j, a position z_j, and a density ρ_j (fractional occupancy between 0 and 1):

$$F(Q)=f(Q)\sum_{j=0}^{\infty}\rho_j e^{iQz_j} \tag{1}$$

where f(Q) is the known form factor, and can be used to include overall scale and Debye-Waller factors as well. An ideal primitive lattice is represented by $\rho_j = 1$ and $z_j = ja$, where a is the perpendicular lattice parameter. Equ 1 reduces to

$$|F(Q)|^2=\frac{|f(Q)|^2}{\sin^2 \frac{1}{2}Qa} \tag{2}$$

demonstrating the $1/\sin^2$ intensity variation characteristic of CTRs. Close to the Bragg peaks at $Q = 2\pi n/a$, the intensity has the asymptotic form $1/|\Delta Q|^2$ that was originally derived by von Laue for the effect of plate morphology on powder diffraction [4], and later for surfaces [5]. It is curious to note that the intensity variation of Equ 2 can be obtained exactly by summing the $1/|\Delta Q|^2$ singularities over all the Bragg peaks, by virtue of the equality

$$\frac{1}{\sin^2 \pi x}=\frac{1}{\pi^2}\sum_{n=-\infty}^{\infty}\frac{1}{(x-n)^2} \ . \tag{3}$$

This has led to an alternative description [6] of extended reflectivity, and CTRs in general, as a superposition of $1/|\Delta Q|^2$ *intensity* tails weighted by the squares of the structure factor at each Bragg peak.

We wish to point out that this equivalence of the two derivations breaks down when the crystal structure is sufficiently complicated. The generalisation of equation (3) works for the case of a primitive lattice of layers with equal weight, AAAA... It also works for an ABABAB... structure, but not for one with a non-uniform layer spacing, such as Si(111), which is illustrated in fig 2. There exist two possible 'ideal' terminations: case 'D' cutting between the double-layers of atoms, and case 'S' splitting the double layer. Fig 2 also shows the extended reflectivity calculated by the two methods. All curves reproduce the correct intensity ratio 2:1:0:1:2 for the (000):(111):(222):(333):(444) Bragg peaks of the diamond lattice. Away from the Bragg peaks they differ, however: in case 'S' there are nodes of zero intensity along the rod profile, but not in 'D'. The superposition of tails calculation generates a curve lying in between, *which is independent of choice of termination*. The differences are not slight: there is almost a factor of 10 difference over most of the region near (222). Very clearly, the profile is sensitive to the choice of termination.

The real Si(111) surface is reconstructed in a way that is described by the well-known 7x7 Dimer-Adatom-Stacking fault (DAS) model [7], shown in a side view in fig 3. This is a relatively complex structure with a reflectivity profile that

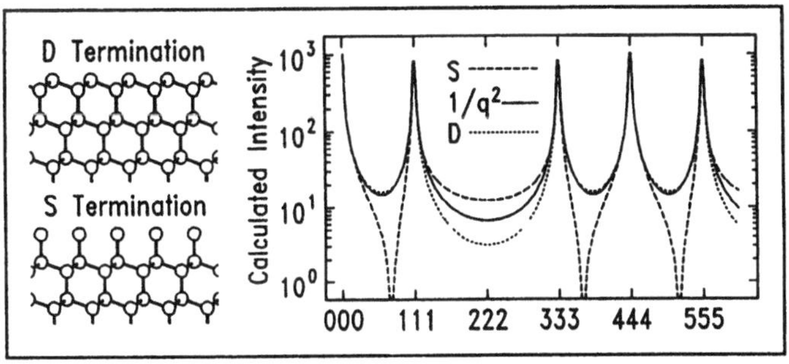

Figure 2. Left: notation for the two possible 'ideal' terminations of the Si lattice by a (111) plane. Right: calculated truncation rod intensity profiles for these ('S' and 'D') with the curve obtained using the sum of intensity tails [6].

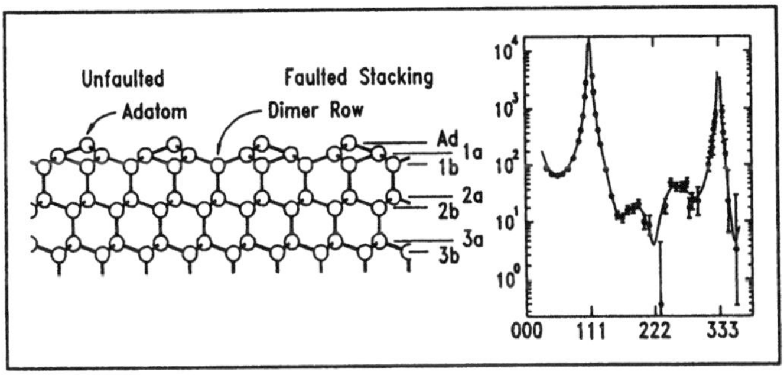

Figure 3. Left: side view of the Dimer Adatom Stacking-fault (DAS) model of the reconstructed Si(111)7x7 surface. Right: measured extended x-ray reflectivity as a function of $Q_\perp$ (bulk Miller index units) and the best fit curve.

would be neither 'S' nor 'D'. Although the general model is no longer questioned, the determination of the structure at the level of atomic positions is still making progress, with recent contributions from Low Energy Electron Diffraction [8] (LEED), Reflection High Energy Electron Diffraction [9] (RHEED) and in-plane x-ray diffraction [10]. As more accurate positions become available, more of the subtle roles of strain and variations in local coordination can be explored [11]. Until now there is no accurate total energy calculation and coordinate optimization of the 7x7 structure because the size of such a project is too large for present day computers, and even this would not answer all the structural questions: the penetration of the strain fields far inside the bulk, for example. The best theoretical prediction of the structure so far comes from a local density approximation (LDA) theory in a 2x2 unit cell [12].

We measured the extended reflectivity of Si(111)7x7 on beamline X16A of the National Synchrotron Light Source (NSLS) at Brookhaven National Laboratory using a UHV x-ray diffractometer [13]. The measurements were made with the detector resolution defined by 2x2mm slits 500mm from the sample and a beam of 2mm in height. At small angle the slits were reduced to 1x1mm and 1mm respectively to avoid running off the sample, and the data were rescaled. The sample was mounted with the diffractometer axis lying in the surface plane, *i.e.* 90° away from the usual setting [13]. In this way a large range of momentum transfers was obtained. The sample was prepared by chemical oxidation using the Shiraki [14] method, followed by resistive heating to about 1200°C to yield a 7x7 LEED pattern.

Data were collected for each point on the reflectivity profile as scans of the diffractometer ω-axis sufficiently far to reach background. The peak was numerically integrated and background subtracted. No trend in the shapes of the peaks was observed; they were always resolution-limited. Data were also taken by scanning in the perpendicular χ direction [13] yielding consistent results. Each intensity point was multiplied by sin2Θ, which is the appropriate Lorentz factor; no polarization correction was needed because of the vertical scattering geometry. A second multiplication by sin2Θ represents the variation of sample area with diffraction angle. Error bars for each point were estimated as a quadrature combination of counting statistics plus a 5% systematic error, found to represent repeated measurements with different alignment, scan direction and slit settings. The full set of measured intensities is shown as the points on figure 3. The 111 and 333 Bragg peaks of Si are clearly seen as divergences of the intensity; no attempt was made to measure closer to these, since resolution and detector saturation effects would start to become important.

The theoretical curve that fits the data in fig 3 is eq(1), modified to include a layerwise Debye-Waller-like 'thickness' term, $\exp(-\frac{1}{2}q^2\zeta_j^2)$, representing the variation of heights within each layer of the structure. The layer occupancies, ρ_j, were taken directly from the DAS model, while the heights, z_j, were adjusted for layers 'Ad' to '3b' in fig 3. The final parameters gave a χ^2 of 0.94. The parameter values and their relation to those obtained by other methods [8,9,12] are discussed elsewhere [15]. There are several oscillations in the curve that arise from the presence of the adatoms (fig 3) and from an expansion of the top layer spacing. The expansion was seen in the LDA calculation [12], but not in LEED [8] or RHEED [9], perhaps because of the limited depth penetration of these electron-based techniques.

NSLS is supported by the U.S. Department of Energy under grant DE-AC012-76CH00016.

References

1. J. Als-Nielsen, in *Handbook on Synchrotron Radiation*, vol. III, ed. D. E. Moncton and G. S. Brown (Elsevier, North-Holland, 1991).
2. M. K. Sanyal, S. K. Sinha, K. G. Huang and B. M. Ocko, Phys. Rev. Lett. **66** 628 (1991); S. K. Sinha, these proceedings (following paper).
3. I. K. Robinson, Phys. Rev. B **33** 3830 (1986).
4. M. von Laue, Annalen d. Physik (folge V) **26** 62 (1936).
5. S. R. Andrews and R. A. Cowley, J. Phys. C **18** 6427 (1985).
6. H. You, these proceedings and private communication.

7. K. Takayanagi, Y. Tanishiro, S. Takahashi and M. Takahashi, Surface Sci. **164** 367 (1985).
8. H. Huang, S. Y. Tong, W. E. Packard and M. B. Webb, Phys. Lett. A **130** 166 (1988).
9. A. Ichimiya, Surf. Sci. **192** L893 (1987).
10. I. K. Robinson, W. K. Waskiewicz, P. H. Fuoss and L. J. Norton, Phys. Rev. B **37** 4325 (1988).
11. M. C. Payne, J. Phys. C **20** L983 (1987); I. K. Robinson, J. Vacuum Science Technology **A6** 1966 (1988).
12. R. D. Meade and D. Vanderbilt, Phys. Rev. B **40** 3905 (1989).
13. P. H. Fuoss and I. K. Robinson, Nucl. Inst. Meth. **222** 171 (1984).
14. A. Ishizaka, N. Nakagawa and Y. Shiraki, in *Proc. 2nd Int. Symp. on MBE and Related Clean Surface Techniques*, 183 (JSAP, Tokyo, 1982).
15. I. K. Robinson and E. Vlieg, submitted to Phys. Rev. B.

Critical Phenomena at Surfaces and Interfaces

R. Lipowsky

Institut für Festkörperforschung, Forschungszentrum Jülich,
Postfach 1913, W-5170 Jülich, Fed. Rep. of Germany

This paper consists of two parts. In the first part, the multitude of critical phenomena which occur at surfaces and interfaces is briefly reviewed. Three types of critical effects are distinguished related (i) to the 2-dimensional character of the interface, (ii) to its morphology, and (iii) to its structural changes at phase transitions in the bulk. In the second part, I focus on *wetting* phenomena which belong to the category (iii). Several wetting processes in three dimensions are discussed: (a) Wetting of a moving interface; (b) Wetting in the 3-dimensional Ising model; and (c) Wetting of an inhomogeneous substrate.

I. Introduction

As shown in many contributions during this conference, both X-rays and neutrons have recently become very powerful probes in order to study the surface of solids and liquids. Such a surface represents the interface between a condensed phase and an inert vapor or 'vacuum'. From a theoretical point of view, one intriguing aspect of interfaces is their *reduced dimensionality.* Indeed, in many cases, the interface can simply be viewed as a planar 2-dimensional system. However, it can also 'bulge' into the third dimension and then attain nonplanar morphologies. In addition, the interface itself has a third dimension which can become mesoscopic as in wetting phenomena; one then has a system which interpolates between two and three dimensions.

This paper consists of two rather different parts. In the first part (Sect. II), I will briefly review the multitude of critical phenomena which can occur at surfaces and interfaces. Three different categories of such phenomena will be distinguished: (i) Critical behavior within the 2-d interface; (ii) Critical effects related to the morphology of interfaces; and (iii) Changes in the interfacial structure as one of the adjacent bulk phases undergoes a phase transition.

In the second part of this paper (Sect. III), I will focus on wetting processes such as, e.g., surface melting or surface-induced order and disorder which belong to the category (iii). In this part, some recent theoretical results for wetting in three dimensions will be discussed: (a) *Wetting of a moving interface* — An example is surface melting of a crystal which slowly evaporates. It is shown that the wetting layer thickens as a result of this motion; (b) *Wetting in the 3-dimensional Ising model* — In this case, thermally excited shape fluctuations of the two interfaces bounding the wetting layer should be important. Until very recently, computer simulations gave no evidence for such fluctuation effects. It is argued here that the interfacial fluctuations observed in the simulations are presumably spikes rather than smooth deformations; and (c) *Wetting of an inhomogeneous substrate* — If one of the phases consists of a disordered solid, the wetting layer experiences a random substrate potential which leads to an increase of the interfacial roughness.

Springer Proceedings in Physics, Vol. 61 **Surface X-Ray and Neutron Scattering**
Eds.: H. Zabel and I.K. Robinson

II. Critical phenomena: a brief overview

A. Critical behavior within the 2–dimensional interface

Usually, the interface between two 3–d phases has a thickness with is set by the scale of the molecules. It then represents a 2–d subsystem which exhibits a variety of phases and associated phase transitions. Consider, e.g., a monolayer of adatoms adsorbed onto the surface of a 3–d liquid or solid. Such a layer often exhibits a 2–d vapor, liquid, and solid phase. In general, the molecules possess internal degrees of freedom which can order and then lead to additional phases. Examples are the liquid crystalline phases of amphiphilic molecules adsorbed at the water–air interface. If the substrate is a crystal, these 2–d layer phases can be commensurate or incommensurate with this substrate.

The multitude of phases leads to a multitude of critical effects. A few examples are: (i) The behavior near the critical point of the 2–d vapor liquid coexistence curve; the associated critical behavior should belong to the universality class of the 2–d Ising model as confirmed in many experiments; (ii) Transitions between commensurate and incommensurate states /1/; and (iii) Wetting phenomena which can occur near 2–d triple points /2/ or can be induced by the steps of a vicinal surface /3/.

The critical effects in examples (ii) and (iii) are governed by the behavior of *domain boundaries* within the 2–d interface (For clarity, 1–d interfaces within 2–d interfaces or surfaces will be called domain boundaries here). The behavior of these domain boundaries is in itself quite interesting.

Consider, for example, a single domain boundary separating two 2–d domains. In equilibrium, the morphology of this domain boundary reflects the underlying *symmetry* of the system /4/. The following cases have theoretically been studied in some detail: (a) *Periodic systems* — In this case, the domain boundary is flat or smooth at $T = 0$. At $T > 0$, it makes transverse excursions which grow with the longitudinal extension, $L_{\|}$, of the domain boundary. More precisely, one has the scaling law $L_{\perp} \sim L_{\|}^{\zeta}$ with the roughness exponent $\zeta = 1/2$; (b) *Quasi–periodic systems* — In this case, the behavior in non–universal and depends on the model parameters /5,6/. If the domain boundary feels an effective Fibonacci potential, it is rough at all T but the roughness exponent ζ is T–dependent and satisfies $\zeta \leq 1/2$; if it feels an effective Harper's potential, the domain boundary undergoes a morphological transition from a smooth state at low T to a rough state at large T; and (c) *Systems with quenched randomness* — In this case, the domain boundary feels a random potential. It then roughens in order to adapt its shape to the random potential and, thus, to minimize its energy. As a result, these systems are characterized by an increased roughness exponent: for random field and random bond systems, one has $\zeta = 1$ (Ref. /7/) and $\zeta = 2/3$ (Ref. /8/) respectively; and (d) *Kinetic roughening* — Away from equilibrium, growth or shrinkage of the domains implies a moving domain boundary. Various models with local deposition or evaporation rules have recently been studied which lead to a kinetic roughening of the domain boundary /9/.

Likewise, the critical effects at wetting in two dimensions have theoretically been studied in great detail. In this case, the system contains two (or several) domain boundaries which attract one another as a result of molecular forces. This direct interaction is, however, strongly renormalized by the shape fluctuations of the domain boundaries. For pure (and periodic) systems, all universality classes for wetting transitions in 2d systems have been identified /10,11/.

B. Critical effects related to the morphology of interfaces

The morphology of 2–d interfaces in three dimensions is completely analogous to the behavior of 1–d domain boundaries discussed in the previous subsection. Thus, the following cases can again be distinguished: (i) *Periodic systems* — The surface of a periodic crystal is smooth at sufficiently low T but undergoes a morphological transition to a rough state at higher T /12/. The roughness is, however, only logarithmic with $L_\perp \sim [\ell n((L_\parallel/a)]^{1/2}$; (ii) *Quasi-periodic crystals* — The surface of an ideal quasicrystal is predicted to be smooth for all T /13,6/. This conclusion is based on renormalization group calculations of interface models; (iii) *Systems with quenched disorder* — In this case, the 2–d interface is presumably always rough with roughness exponent $\zeta > 0$. This behavior should also apply to an interface in a random quasicrystal /13/; and (iv) *Kinetic roughening* — Local deposition rules again lead to an increased roughness of the moving interface and typically to an increased value of ζ. In addition, various morphological transitions between two different rough states of the interface have been predicted /9/.

Another aspect of the morphology which is governed by fluctuations is the singular behavior of the equilibrium crystal shape which occurs near the edge of a facet. At such an edge, the facet often meets a rounded part composed of terraces and steps. Let $z(x)$ describe this shape: the facet is given by $z(x) = 0$ for $x \le 0$ and the edge of this facet is at $x = 0$. For a periodic crystal, the rounded part is then described by $z(x) \sim x^\lambda$ with $\lambda = 3/2$ (Ref. /14/). In general, scaling arguments lead to $\lambda = 1/\tau$ (Ref. /13/) with $\tau = 1/\zeta$ for $\zeta \le 1/2$ and $\tau = 2(1-\zeta)/\zeta$ for $\zeta \ge 1/2$, where ζ is the roughness exponent of the 1–d ledges or steps separating the terraces of the rounded part. This scaling relation nicely illustrates the interplay of 2–d and 3–d properties of the interface.

C. Interfacial structure at phase transformations in the bulk

When one of the two bulk phases adjacent to the interface undergoes a phase transition, the interface aquires characteristic density profiles. To be more specific, let us consider a crystal–vapor interface, i.e., the surface of a crystal. The bulk crystal may undergo a variety of *continuous* phase transitions. If this transition belongs to the universality class of the 3–d Ising model, one can distinguish three types of surface behavior /15/: (i) If the coupling constant, J_1, between two "spins" within the surface is smaller than the coupling constant, J, for two "spins" in the bulk, the system undergoes an 'ordinary transition' at the bulk critical temperature $T = T_c$ at which both bulk and surface become ordered simultaneously; (ii) If J_1 is large compared to J, the surface undergoes a 'surface transition' at $T_{c1} > T_c$ while the bulk undergoes an 'extraordinary transition' at $T = T_c$; and (iii) Finally, for a certain intermediate ratio J_1/J, bulk and surface simultaneously undergo the so–called 'special transition'.

Now, consider again the interface between a crystalline phase, α, and a vapor phase, γ, but assume that the crystal undergoes a *discontinuous* phase transition between a disordered phase at high temperature and several odered phases at low temperature. Then, a third phase, β, appears which may prefer to go into the $(\alpha\gamma)$ interface and then forms a thin wetting layer. Two cases may be distinguished: as the transition temperature is approached (i) from below or (ii) from above, a thin layer (i) of the disordered phase or (ii) of one of the ordered phases can be induced by the surface. /16,17/ Alternatively, if both bulk phases α and γ are in chemical equilibrium, wetting by the β phase may occur as one approaches a triple point at which all three phases α, β and γ coexist.

When the $(\alpha\gamma)$ interface contains a wetting layer of β phase, it splits up into an $(\alpha\beta)$ and a $(\beta\gamma)$ interface. The separation, $\ell \ge 0$, of these two interfaces is equal to the thickness of the wetting layer. Likewise, the excess free energy, $V(\ell)$, of the layer can be regarded as an effective interaction between these interfaces.

In the absence of interfacial fluctuations, the interaction $V(\ell)$ has the general form /4/

$$V(\ell) = H\ell + V_{DI}(\ell)\,. \tag{1}$$

The pressure–like variable H is proportional to the difference, $f_\beta - f_\alpha$, of the bulk free energies of the metastable β and the stable α phase. The direct interaction, $V_{DI}(\ell)$, reflects the underlying microscopic forces such as van der Waals, electrostatic, or structural forces.

The interaction $V(\ell)$ as given by (1) is, in general, renormalized by shape fluctuations of the interfaces even though this renormalization is less effective in $d = 3$ than in $d = 2$. The influence of interfacial fluctuations is studied most conveniently within the framework of effective models for the interfacial separation $\ell \geq 0$. For wetting in three dimensions, these models have the generic form /4,17/

$$\mathcal{H}\{\ell\} = \int d^2x \left\{\frac{1}{2}\tilde{\Sigma}(\nabla\ell)^2 + V[\ell(\vec{x})]\right\} \tag{2}$$

where $\tilde{\Sigma}$ is an appropriate interfacial stiffness and $V(\ell)$ is given by (1). A hard wall potential at $\ell = 0$ which ensures $\ell \geq 0$ is implicitly assumed. In what follows, I will study three different situations which correspond to three different interactions, $V_{DI}(\ell)$.

III. Wetting phenomena in three dimensions: some recent theoretical results

A. Wetting away from equilibrium: a moving wetting layer is thicker

First, consider an $(\alpha\gamma)$ interface which moves with constant velocity. An example is a crystal in ulta–high vacuum which slowly evaporates. Now, assume that this interface contains a thin wetting layer. The question is: does the motion of the interface affect the thickness of the layer?

This question can be addressed in the framework of Landau models or density functional theories which lead to density profiles, $M(z,t)$, where z is the coordinate perpendicular to the $(\alpha\gamma)$ interface and t is the time. Let us assume that the motion of the interface is not limited by diffusion. One may then consider a simple relaxational dynamics for the densities /18-20/. This leads to solutions $M(z,t) = f(z - vt)$ which move with constant velocity v.

The behavior of the densities can be used in order to construct the interaction $V(\ell)$ acting between the $(\alpha\beta)$ and the $(\beta\gamma)$ interface. As a result, one finds a shift of the pressure–like variable H and the direct interaction /19/

$$V_{DI}(\ell) = B\exp(-\ell/\ell_1) + C\exp(-\ell/\ell_2) \tag{3}$$

where the two length scales are given by

$$\ell_1 = \xi_\beta/(-x + \sqrt{1+x^2}) \qquad \text{and} \qquad \ell_2 = \xi_\beta/2\sqrt{1+x^2} \tag{4}$$

with $x \equiv v/v_\beta$ where v_β is a velocity scale which depends on the mobility within the β layer. The length scale ξ_β is the bulk correlation length within the β phase. For $x = v/v_\beta = 0$, one has $\ell_1 = \xi_\beta = 2\ell_2$. In general, the scale ℓ_1 increases with increasing velocity. If one ignores the possible effects of shape fluctuations, the expectation value for ℓ is determined by the minimum of $V_{DI}(\ell) + H\ell$. It then follows that the motion of the interface leads to a thickening of the wetting layer /19/.

B. Wetting in the 3–dimensional Ising model: is it dominated by spikes?

Next, consider the Ising model on a simple cubic lattice with a (100) surface. The "spins" interact with nearest–neighbor couplings, J, and are subject to a short–ranged surface field. This model exhibits critical wetting transitions which have extensively been studied in Monte Carlo simulations /21/. However, the critical behavior of these transitions is not well–understood. Until very recently, the simulations within the Ising model were thought to be consistent with mean–field theory. In contrast, linearized renormalization group calculations of effective interface models with

$$V_{DI}(\ell) = B\exp(-\ell/\ell_0) + C\exp(-2\ell/\ell_0) \tag{5}$$

predicted nonuniversal critical behavior depending on the parameter $\omega = T/4\pi\tilde{\Sigma}\ell_0^2$ where $\tilde{\Sigma}$ is the interfacial stiffness as in (2). /22,23/

In order to address this controversy, we have studied the solid–on–solid (SOS) limit of the 3–d Ising model by Monte Carlo simulations /24/. A detailed analysis of the MC data revealed that the SOS model belongs to the same universality class as the Gaussian model given by (2) provided (i) one accepts the results of linear renormalization, and (ii) one makes a proper identification of the model parameters. The reduced stiffness of the SOS model has the value $\tilde{\Sigma}/T = c_1(J/Ta)^2$ with $c_1 \simeq 10.4$, and the direct interaction has the form (5) with $\ell_0 = \xi = c_2 aT/J$ and $c_2 \simeq 0.175$. As a consequence, the parameter ω for the SOS limit of the 3–d Ising model has the universal value $\omega \simeq 1/4$. /24/

The direct interaction (5) for the Ising model can be derived from mean–field theory for the order parameter density /22,25,26/. In this derivation, the length scale ℓ_0 is identical with the bulk correlation length ξ_β within the β layer as follows from (3) and (4) for $x = v/v_\beta = 0$. The SOS model does not include any bubbles or interfacial overhangs and thus does not contain information about the bulk correlation length. Therefore, the length scale ξ cannot be identical with the bulk correlation length. However, this length scale has another rather direct interpretation in terms of interfacial *spikes*. Thus, consider two planar interfaces at separation ℓ and introduce an interfacial fluctuation which consists simply of a column of reversed spins bridging the gap between the two interfaces. Such a fluctuation has an energy $\Delta E = 8J\ell/a$ and thus a Boltzmann weight $\sim \exp(-8J\ell/aT)$. Therefore, ignoring entropic effects, these spikes are governed by the length scale $\xi_{sp} = 0.125aT/J$ which is comparable to the scale ξ determined in the simulations.

Thus, one may distinguish two different types of shape fluctuations of the interface: (i) smooth deformations of the mean–field profile, and (ii) abrupt spikes in the form of thin fingers. In both cases, the critical behavior resulting from the shape fluctuations can be described by an effective Gaussian model as in (2) with a direct interaction as in (5) *but with different length scales* ℓ_0. The question now is: which type of shape fluctuation is *typical* for critical wetting in the 3–d Ising model?

For $J/T = 0.35$, ξ as found in our simulations is about twice as large as the generally accepted value for the bulk correlation length ξ_β. /24/ In this case, spikes should dominate and $\omega \simeq 1/4$. This conclusion is consistent with a recent analysis of the MC data for the 3–d Ising model /27/. This analysis is based on the amplitude ratio for the interfacial susceptibility χ_1 and leads to the estimate $\omega \simeq 0.3$ which is rather close to the value $\omega \simeq 1/4$ obtained for the SOS model.

C. Wetting of an inhomogeneous substrate: lateral disorder translates into interfacial roughness

Finally, consider wetting of a smooth solid substrate which has an inhomogeneous composition and thus exerts an effective interaction with a random component. /28/ The corresponding direct interaction is taken to be

$$V_{DI}(\ell) = V_R(\ell) - (W + W_r)v(\ell)\,. \tag{6}$$

The first term, $V_R(\ell)$, and the second term, $-Wv(\ell)$ with $W \geq 0$, describe the repulsive and the attractive part of V_{DI}, respectively. The third term, $-W_r v(\ell)$, corresponds to the random component of V_{DI} where W_r is a quenched random variable with zero mean value, $\overline{W}_r = 0$. The potential $v(\ell)$ is taken to behave as $v(\ell) \sim 1/\ell^s$ for large ℓ; (non-retarded) van der Waals forces are described by $s = 2$.

For complete wetting with $W = 0$, the $(\alpha\beta)$ interface unbinds from the substrate as the pressure-like variable H in (1) goes to zero. Alternatively, this interface undergoes a critical wetting transition at $(\alpha\beta)$ coexistence with $H = 0$ in the limit of zero W. For a relatively thick wetting layer, the interfacial fluctuations are now characterized by $L_\perp \sim L_\parallel^\zeta$ with $\zeta = 1/(2+s)$ which arises from the random component of the substrate potential. /29/ This increased roughness affects the mean separation $< \ell >$ of the $(\alpha\beta)$ interface from the substrate *provided* $V_R(\ell)$ decays faster than $1/\ell^\tau$ with $\tau = 2(1+s)$. In the latter situation, one has $< \ell > \sim 1/H^{\psi_c}$ with $\psi_c = 1/(3+2s)$ for complete wetting and $< \ell > \sim 1/W^\psi$ with $\psi = 1/(2+s)$ for critical wetting.

Acknowledgements

I thank Michael E. Fisher for a helpful discussion about interfacial spikes, and Joanna Cook, Willi Fenzl, Stefan Grotehans, Frank Jülicher, Götz Schmidt, Jörn Sonnenburg, Dietrich Wolf and Joachim Wuttke for stimulating interactions. I also thank the organizers of this conference for their invitation.

References

/1/ For a review, see M.E. Fisher, J. Chem. Soc. Far. Trans. II **82**, 1569 (1986).

/2/ An example is provided by edge melting which leads to a singular contribution $\sim (T_t - T)^{-\alpha}$ in the surface specific heat. For adatoms interacting via van der Waals forces, one has $\alpha = 4/3$, see R. Lipowsky, Phys. Rev. B**32**, 1731 (1985). Experiments on Ne monolayers by D.M. Zhu, D. Pengra, and J.G. Dash, Phys. Rev. B**37**, 5586 (1988) lead to the estimate $\alpha \simeq 5/4$.

/3/ This has recently been studied by Monte Carlo simulations, see K. Binder, in *Kinetics of ordering and growth of surfaces* ed. by M. Lagally (Plenum, New York 1990).

/4/ For reviews, see R. Lipowsky, in *Random fluctuations and patterns growth*, ed. by H.E. Stanley and N. Ostrowsky (Kluwer Academic, Dordrecht 1988) p. 227; G. Forgacs, R. Lipowsky, and T.M. Nieuwenhuizen, to appear in *Phase transitions and Critical Phenomena*, ed. by C. Domb and J. Lebowitz.

/5/ C. Henley and R. Lipowsky, Phys. Rev. Lett. **59**, 1679 (1987); A. Garg and D. Levine, Phys. Rev. Lett. **59**, 1683 (1987).

/6/ R. Lipowsky, in *Fundamental problems in statistical mechanics VII*, ed. by H. van Beijeren (North–Holland, Amsterdam 1990) p. 139, and references therein.

/7/ J. Villain, J. Phys. (Paris) Lett. **43**, L551 (1982); G. Grinstein and S.-K. Ma, Phys. Rev. B**38**, 2588 (1983).

/8/ D.A. Huse, C.L. Henley, and D.S. Fisher, Phys. Rev. Lett. **55**, 2924 (1985); M. Kardar, Phys. Rev. Lett. **55**, 2923 (1985).

/9/ For reviews, see D.E. Wolf in *Kinetics of ordering and growth at surfaces*, ed. by M. Lagally (Plenum Press, New York 1990); and J. Krug and H. Spohn, in *Solids far from equilibrium: Growth, morphology and defeats*, ed. by C. Godreche (Cambridge University Press, 1991).

/10/ R. Lipowsky and T.M. Nieuwenhuizen, J. Phys. **A21**, L89 (1988); R.K.P. Zia, R. Lipowsky, and D.M. Kroll, Am. J. Phys. **56**, 160 (1988); F. Jülicher, R. Lipowsky, and H. Müller–Krumbhaar, Europhys. Lett. **11**, 657 (1990); H. Spohn, Europhys. Lett. **14**, 689 (1991).

/11/ Wetting in 2–d quasiperiodic systems and in 2–d random bond systems has recently been studied by G. Schmidt and R. Lipowsky, to be published, and by J. Wuttke, and R. Lipowsky, submitted to Phys. Rev. B.

/12/ A recent review is in H. van Beijeren and I. Nolden, in *Structure and Dynamics of Surfaces II*, ed. by W. Schommers and P. van Blanckenhagen (Springer–Verlag, 1987).

/13/ Lipowsky and Henley, Phys. Rev. Lett. **60**, 2394 (1988).

/14/ C. Rottmann, M. Wortis, J.C. Heyraud, and J.J. Metois, Phys. Rev. Lett. **52**, 1009 (1984).

/15/ For reviews, see K. Binder, in *Phase transitions and critical phenomena*, Vol. 8, ed. by C. Domb and J. Lebowitz (Academic Press, London 1983); H. Wagner, in *Applications of field theory to statistical mechanics*, ed. by L. Garrido (Springer, Berlin 1985); H.W. Diehl, in *Phase transitions and critical phenomena*, Vol. 10, ed. by C. Domb and J. Lebowitz (Academic Press, London 1986).

/16/ R. Lipowsky, Phys. Rev. Lett. **49**, 1575 (1982); R. Lipowsky and W. Speth, Phys. Rev. **B28**, 3983 (1983); for reviews, see R. Lipowsky, J. Appl. Phys. **55**, 2485 (1984); and in *Magnetic properties of low–dimensional systems II*, ed. by L.M. Falicov, F. Mejia–Lira, and J.L. Moran–Lopez (Springer–Verlag, Berlin 1990) p. 158.

/17/ Surface–induced order in *fcc* alloys has recently been studied by W. Schweika, K. Binder, and D.P. Landau, Phys. Rev. Lett. **65**, 3321 (1990).

/18/ T. Meister and H. Müller–Krumbhaar, Z. Physik **B55**, 111 (1984).

/19/ H. Löwen and R. Lipowsky, Phys. Rev. **B43**, 3507 (1991).

/20/ Diffusion–limited growth of wetting layers is studied in R. Lipowsky and D.A. Huse, Phys. Rev. Lett. **57**, 353 (1986).

/21/ K. Binder and D.P. Landau, Phys. Rev. **B37**, 1745 (1988).

/22/ R. Lipowsky, D.M. Kroll, and R.K.P. Zia, Phys. Rev. **B27**, 4499 (1983) used a variational method which is equivalent to linear renormalisation. Their parameter τ corresponds to $[4\pi\omega]^{1/2}$.

/23/ E. Brezin, B. Halperin, and S. Leibler, Phys. Rev. Lett. **50**, 1387 (1983); D.S. Fisher and D.A. Huse, Phys. Rev. **B32**, 247 (1985).

/24/ G. Gompper, D.M. Kroll, and R. Lipowsky, Phys. Rev. **B42**, 961 (1990).

/25/ E. Brezin, B. Halperin, and S. Leibler, J. Phys. (Paris) **44**, 775 (1983).

/26/ M.E. Fisher and A.J. Jin, University of Maryland, preprint.

/27/ A.O. Parry, R. Evans, and K. Binder, University of Mainz, preprint.

/28/ This situation has to be distinguished from wetting in 3–d random systems, see R. Lipowsky and M.E. Fisher, Phys. Rev. Lett. **56**, 472 (1986), and from wetting of a rough solid substrate, see M. Kardar and J. Indekeu, Europhys. Lett. **12**, 161 (1990); H. Li and M. Kardar, Phys. Rev. **B42**, 6546 (1990).

/29/ For a $d_{||}$–dimensional interface in $d = d_{||} + 1$, one has $\zeta = (5-d)/2(2+s)$ which exceeds the thermally–excited roughness for $s < s^*$ with $s^* = (d-1)/(3-d)$ for $1 < d \leq 3$ and $s^* = \infty$ for $3 < d \leq 5$. If $s < s^*$ and $V_R(\ell) \ll 1/\ell^\tau$ with $\tau = 2(d-1+2s)/(5-d)$, complete and critical wetting are characterized by $\psi = (5-d)/(d+3+4s)$ and $\psi = (5-d)/(d-1)(2+s)$, respectively.

Surface-Induced Order Observed on a $Cu_3Au(001)$ Surface

K.S. Liang[1], *H.H. Hung*[1], *S.L. Chang*[1], *Z. Fu*[1], *S.C. Moss*[2], and *K. Oshima*[3]

[1]Exxon Research and Engineering Co., Annandale, NJ 08801, USA

[2]University of Houston, Houston, Texas, USA

[3]Tsukuba University, Tsukuba, Ibaraki, Japan

1. Introduction

The possibility of surface effects on the first-order transition was first proposed by Lipowsky [1]. At temperatures away from the bulk transition, the surface can exhibit disordering (if approached from below the transition) or ordering (if approached from above the transition) phenomena. So far most experimental studies on surface-induced transition have been centered on the surfaces of a classical binary alloy, Cu_3Au [2-4]. In the case of the Cu_3Au (001) surface, previous LEED and x-ray studies [2,3] suggested that a surface-induced disorder took place below the bulk transition. These results also indicated that the surface transition observed was consistent with a critical wetting. Recently, we studied the surface of a 2% Au-rich Cu_3Au sample ($T_c \sim 376\ ^oC$) and found a very strong surface segregation effect [5]. Such effect produces a segregated Cu_3Au layer at the surface with its order-disorder temperature $\sim 390\ ^oC$ (i.e., close to the bulk T_c). We therefore re-investigated a second crystal, of which the stoichiometry is Cu_3Au to our experimental accuracy. In contrast to previous studies [2,3], our results showed a *surface-induced order* on this sample.

2. Experimental

The surface of the mechanically polished sample was oriented to within 0.3^o of the (001) orientation. The stoichiometry of the sample was determined by measuring the lattice constant (a_o = 3.752 Å) and the bulk transition temperature ($T_c = 390 \pm 2\ ^oC$) in a nitrogen filled oven. X-ray measurements were performed on Exxon beamline X10A at the National Synchrotron Light Source using a z-axis type surface scattering spectrometer. Extensive surface cleaning was performed through cycles of sputtering with 0.5 KV Ar^+ gas and long annealing.

Most measurements were taken with a fixed incident angle of x-rays near the critical angle of Cu_3Au (0.31^o for 11,500 eV x-rays). Detailed x-ray scans of the surface structure were carried out in the reciprocal space along the (100)-(101) rod. The sample temperature was calibrated with respect to the above bulk transition temperature by measuring the (001) Bragg intensity . The longitudinal resolution of the diffracted beam was set by slits to be .0095 $Å^{-1}$ for both in-plane and out of plane scans.

3. Experimental results and discussions

Figure 1 shows typical x-ray scans taken around the (101) Bragg peak at temperatures *above the bulk transition temperature.* Scattering profiles shown in the figure were scans taken in the [100] direction. The broad diffuse background shown is associated with the short-range order from the bulk. The most interesting observation is the sharp component (coherent length ~ 350 Å) shown on top of the diffuse background. This is a direct evidence that an ordered surface layer is present on the (001) surface of the disordered bulk Cu_3Au.

Our x-ray scans taken around the (100) Bragg peak showed similar results; i.e. an ordered surface layer was revealed by a sharp in-plane (100) surface peak was at temperatures above the bulk transition. In Figure 2, the measured peak intensity of the (100) reflection is plotted as a function of temperature, together with that of the (001) bulk peak. Following the heating cycle from an ordered state, the bulk (001) peak intensity dropped 4 orders of magnitude at the bulk transition temperature. While the intensity of the (100) surface peak dropped by a factor of about 50 at T_C, the peak remains to be very sharp (coherent length ~ 300 Å) at temperatures above T_C. Eventually, the peak merged with the diffuse short-range-order background at 20 °C above T_C. It should be mentioned that the observed scattering profiles of the (100) peak observed between T_C and $T_C \pm 20$ °C are reversible. Empirically, the observed profiles can be fitted with a Gaussian plus Lorentzian lineshape. We attribute the Gaussian component to the presence of an ordered surface layer.

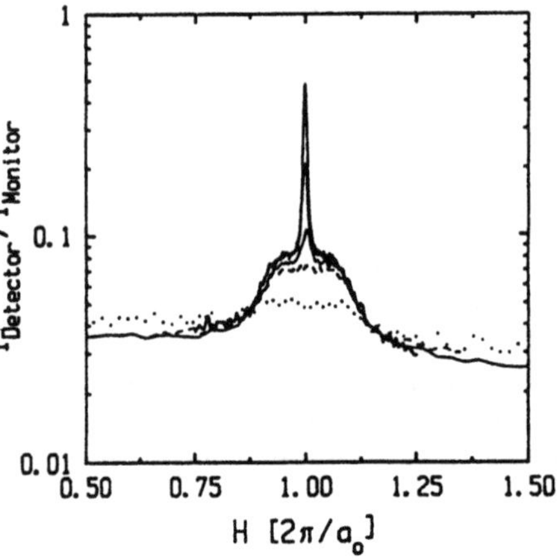

Fig. 1. The scattering profiles around the (101) peak measured in the [100] direction at temperatures near and above the bulk transition temperature.

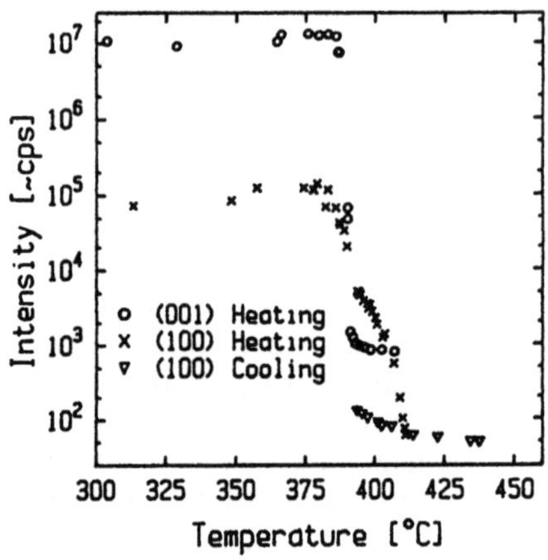

Fig. 2. The peak intensity of bulk (001) and surface (100) reflections as function of temperature. The temperature is calibrated to the bulk transition temperature of 390 °C.

In the cooling cycle, the sample was first annealed at 500 °C into a disordered state. As the temperature was lowered, surface order was again observed at temperatures far above T_c. However, the in-plane measured coherent length of the surface ordered layer was found to be smaller in this case than that observed in the heating cycle from a well ordered state. The difference observed between heating and cooling cycles suggests that impurity pinning might inhibit the lateral growth of the surface ordered layer as the sample is approached from a disordered state.

In summary, our x-ray studies show that the (001) surface of Cu_3Au remains order at temperatures above the bulk order-disorder transition temperature. The behavior observed for the surface-induced ordering transition is consistent with a wetting phenomenon [1]. Our studies clearly show that sample compositions play a very crucial role on the surface order-disorder transition.

References

1. R. Lipowsky, Phys. Rev. Lett. 49, 1575 (1982); J. Appl. Phys. 55, 2485 (1984).
2. S.F. Alvarado, M. Campagna, A. Fattah, and W. Uelhoff, Z. Phys. B 66, 103 (1987).
3. H. Dosch, L. Mailander, A. Lied, J. Peisl, F. Grey, R. L. Johnson, and S. Krummacher, Phys. Rev. Lett. 60, 2382 (1988).
4. X-M. Zhu, H. Zabel. I.K. Robinson, E. Vlieg, J.A. Dura, and C.P. Flynn, Phys. Rev. Lett. 65, 2692 (1990).
5. H.H. Hung, S.B. Rivers, K.S. Liang, W.N. Unertl, and K.L. D'Amico, to be published.

Thermal Dynamics of (110) fcc Metal Surfaces

K. Kern

IGV-KFA Jülich, Postfach 1913, W-5170 Jülich, Fed.Rep.of Germany
and Institut de Physique Experimentale, EPF Lausanne,
PH-Ecublens, CH-1015 Lausanne, Switzerland

Abstract We discuss He- and x-ray diffraction data on the thermal behavior of the clean Cu(110) and Pt(110)(1x2) surface. The unreconstructed Cu(110) surface exhibits already at temperatures of $T \simeq 0.4\ T_M$ a pronounced thermal anomaly, attributed to an enhanced surface anharmonicity. Upon further increase of the temperature around $T_R \simeq 0.8\ T_M$ steps appear spontaneously and the surface roughens. On the reconstructed Pt(110)(1x2) surface, monoatomic steps are formed spontaneously already at temperatures $T_R \simeq 0.53\ T_M$.

The loss of vertical translational invariance at the surface changes the local force field with respect to the bulk forces. This charge redistribution is responsible for noticeable interlayer relaxations in the near surface region of most crystals. In some cases the changes in the force field can even cause a lateral atomic rearrangement in the surface plane, the surface reconstructs into a phase with new symmetry. The restructuring can either occur spontaneously or be activated by small amounts of adsorbates.

Restructuring can also be caused by simple temperature variations. As the temperature rises the lattice vibrational amplitude increases and the anharmonic terms in the interaction potential become important. Due to the reduced number of nearest neighbors in the surface the mean-square amplitude of the surface atoms is much larger than in the bulk, driving the surface to disorder at elevated temperatures. In particular crystal planes with an open structure are subject to surface roughening /1/ and surface melting phase transitions /2/.

The concept of a roughening transition has been developed by Burton, Cabrera, and Frank /3/. It plays a crucial role in the understanding of the crystal growth and the equilibrium shape of crystal surfaces. On an atomic scale the roughening of a clean smooth surface is defined as the proliferation of atomic steps by thermal generation at the roughening temperature T_R at which the free energy for creation of a step becomes zero.

X-ray and He-atom diffraction has been used to study the thermal dynamics of the (110) surface of face centered cubic metals. These metal surfaces separate in two different clases. The first class, including the 3d-elements Cu, Ni and the 4d-elemenents Rh, Pd and Ag, have a nonreconstructed (1x1) ground state for the clean surface, i.e. they keep the bulk symmetry. The second class of fcc metals, including the 5d-elements Ir, Pt, and Au, exhibits a reconstructed (1x2) ground state with missing row geometry. As representatives of these two different clases we will discuss here result for Cu and Pt, respectively.

More than ten years ago it had been noticed that the intensities in the photoemission spectra taken from Cu(110) decrease dramatically with temperature above ~ 500 K /4/. Similar effects have been seen recently in low-energy ion scattering /5/, in X-ray diffraction /6/ and in thermal He scattering /7,8/. The dramatic intensity decrease observed in all cases above 450-500 K could not be accounted for by simple Debye-Waller effects. While Lapujoulade et al. /7/ and Fauster et al. /5/ proposed as explanation either anharmonic effects or some kind of disorder, Mochrie /6/ concluded catagorically - without qualitative additional evidence - that he was observing the roughening transition. He even tentatively identified the temperature at which "the intensity has fallen essentially to zero" (870 K) with T_R. In a later study Ocko and Mochrie /6b/ realized that the sample used in the original experiments of Mochrie had a large miscut of about 0.8^o and that such miscut samples did actually exhibit a reversible facetting transition with the coexistence of large flat (110) terraces and inclined rough regions (clustered steps) at low temperatures. Recently Robinson et al. /19/ observed a similar behavior for Ag(110), demonstrating that the roughening temperature is a rapid function of the surface miscut (decreasing T_R with increasing miscut).

A He specular intensity measurement on an almost perfect Cu(110) crystal with a miscut of less than 0.1^o versus temperature performed in our laboratory shows that also above 870 K the intensity continues to drop (at 1000 K it is one order of magnitude lower) and that there is no sign of saturation even above 1000 K. Whether the intensity becomes "essentially zero" appears to depend on the dynamical range of the instrument, and is not a criterion for the choice of value of T_R. Zeppenfeld et al. /8/ have analyzed in detail the energy and angular distribution of the scattered He atoms in the whole temperature range up to 1000 K, and extended these measurements recently up to 1200K. This analysis showed that the anomaly at $T \simeq 0.4\ T_M$ is not accounted for by structural changes of the surface but due to an anomalous increase of the mean square displacement of the surface atoms above that temperature, which can be ascribed to an enhanced surface anharmonicity. A recent EELS study /9/ of the temperature dependence of the surface phonons has quantified this effect; according to this experiment the surface anharmonicity of the Cu(110) surface is enhanced about 5 times over that of bulk copper.

Upon further increase of temperature isolated defects start to appear at about 70-75% of the melting temperature and eventually around $T_R \simeq 1070\ K \simeq 0.8\ T_M$ steps appear spontaneously and the surface roughens. The measured roughening temperature of 1070 K coincides with the temperature at which the anisotropy in the activation energy for surface diffusion of adatoms on the Cu(110) surface vanishes /10/. Above this temperature the mass transport is isotropic, consistent with 2D step diffusion.

In theoretical studies of the reconstructed fcc(110) surfaces it has been shown that the (1x2) missing row configuration is indeed only marginally stable with respect to the "higher" missing row states (1x3, 1x4, ..., 1xn). The energy difference between any of the (1xn) phases of Au(110) has been calculated to be less than 10 meV per atom /11/. Based on this ground it has been argued by several authors that the missing row configuration should be thermally unstable with respect to the formation of (111) microfacets, giving rise to a "rough" surface at temperatures well below T_R of the non reconstructed (110) surfaces. While Villain and Vilfan /12/ have predicted a succession of two transitions, an Ising-like order-disorder transition at $\sim 0.50\ T_M$ spontaneous proliferation of antiphase Ising-defects, followed by a roughening transition at $\sim 0.57\ T_M$ (onset of (111) micro facetting generating single height steps), Levi and Touzani /13/ have found no evidence for an Ising-like transitions but predicted a direct roughening transition.

In a recent x-ray diffraction experiment Robinson, Vlieg and Kern have studied the thermal behavior of the reconstructed Pt(110) surface /14/. The experimentally observed half order diffraction peaks have two characteristics: they are broad in the [001] direction but sharp in the orthogonal [1$\bar{1}$0] direction and always displaced slightly from the exact half order position along [001]. The uniaxial broadening and shift implies disorder in one direction only, i.e. must be associated with line defects oriented perpendicular to the [001] direction. An identical pattern of uniaxially shifted and broadened half order diffraction peaks was observed earlier by Robinson et al. /15/ for the Au(110)(1x2) surface and can be explained conclusively in terms of randomly distributed single height steps on the surface. It was further demonstrated that the peak shift of the half order spots is exclusively related to the density of these monatomic steps while Ising-like defects would only result in a symmetric peak broadening.

The basic result of this study is that the half-order diffraction spots not only gradually disappear with rising temperature but also show a peak shift which is temperature dependent. Above $T > T_R = 1080$ K = 0.53 T_M the half order diffraction peak broadens and shifts dramatically and completely reversible indicating the simultaneous proliferation of antiphase Ising-defects and of steps. No separation in the proliferation of steps and Ising-defects were detected within the accuracy of the experiment. This and the observation of Ising-exponents for the transition has caused some controversial discussion /16-18/.

References

/1/ H. van Beijeren and I. Nolden; in Structure and Dynamics of Surfaces II (Springer, Berlin, 1986), p.259

/2/ J.G. Dash; Contemp. Phys. 30, 89 (1989)

/3/ W.K. Burton, N. Cabrera, F.C. Frank; Philos. Trans. Roy. Soc. London 243 A, 299 (1951)

/4/ R.S. Williams et al.; Phys. Rev. Lett. 39, 302 (1977)

/5/ Th. Fauster et al.; Surf. Sci. 189/190, 610 (1987)

/6/ S.G.J. Mochrie; Phys. Rev. Lett. 59,304 (1987), B.M. Ocko and S.G.J. Mochrie; Phys. Rev. B38, 7378 (1988)

/7/ J. Lapujoulade, J. Perreau and A. Kara; Surf. Sci. 129,59 (1983)

/8/ P. Zeppenfeld et al.; Phys. Rev. Lett. 62, 63 (1989); K. Kern et al., to be published

/9/ A.P. Baddorf and E.W. Plummer; Phys. Rev. Lett. 66, 2770 (1991)

/10/ H.P. Bonzel and E. Latta; Surf. Sci. 76, 275 (1978)

/11/ M. Garofalo, E. Tosatti, F. Ercolessi; Surf. Sci. 188, 324 (1987)

/12/ J. Villain and I Vilfan; Surf. Sci. 199, 165 (1988)

/13/ A.C. Levi and M. Touzani; Surf. Sci. 218, 223 (1989)

/14/ I.K. Robinson, E. Vlieg, K. Kern; Phys. Rev. Lett. 63, 2578 (1989)

/15/ I.K. Robinson, Y. Kuk, L.C. Feldman; Phys. Rev. B29, 4762 (1984)

/16/ I. Vilfan and J. Villain; Phys. Rev. Lett. 65, 1830 (1990)

/17/ I.K. Robinson, E. Vlieg, and K. Kern; Phys. Rev. Lett. 65, 1831 (1990)

/18/ M. den Nijs; Phys. Rev. Lett. 66, 907 (1991)

/19/ I.K. Robinson, E. Vlieg, H. Hornis, E.H. Conrad; to be published, see also this proceeding

Facet Coexistence in the Roughening Transition of Ag(110)

H.G. Hörnis[1], *E.H. Conrad*[1], *E. Vlieg*[2;a], *and I.K. Robinson*[2]

[1]Department of Physics and Astronomy, University of Missouri, Columbia, MO 65211, USA
[2]AT&T Bell Laboratories, Murray Hill, NJ 07974, USA
[a]Permanent address: FOM Institute, Kruislaan 407, 1098 SJ Amsterdam, Netherlands

Abstract. We have studied the thermal behavior of the Ag(110) surface by synchrotron x-ray diffraction. In-plane diffraction data ($Q_z \approx 0$) agrees with earlier work and can be fit to a power law form. Out-of-plane data ($Q_{z,max}$=0.7), however, indicates the existence of two coexisting phases below the roughening transition: flat (110) oriented regions separated by inclined rough regions. The relative coverage of these two phases is found to depend on the temperature. Thus, the roughening process can be viewed as a continuous replacement of the flat faceted regions by the rough phase. Using the Wulff construction, we are able to describe the temperature dependence of the relative phase concentrations and extrapolate to an estimate of the Ag(110) roughening temperature.

Introduction. There is a great deal of interest in surface roughening, primarily because of its fundamental relationship to Equilibrium Crystal Shapes (ECS). The thermal evolution of the ECS of small Au and Pb crystals have been studied [1]. One of the results of those experiments was that close to the melting point only (100) and (111) facets persisted; all other orientations being rough by this temperature. Roughening of a flat surface has been predicted for several systems. Chui *et. al.*, showed that the Solid-on-Solid (SOS) model has a phase transition leading to a high temperature phase with a logarithmic height divergence [2],

$$\langle (h(r)-h(0))^2 \rangle \propto \ln(r), \qquad (1)$$

where h(r) is the height of the surface a distance r from an arbitrary origin. Roughening transitions have been studied using x-ray diffraction [3,4], helium atom scattering [5,6] and LEED [7] for various metal surfaces. The expected diffraction lineshape has been calculated to have a power law form [4,6]

$$I(q_{||}, q_z) \propto \frac{1}{\sin^2(q_z/2)} \left| q_{||}^2 \right|^{\eta(T,q_z)/2 - 1}, \qquad (2)$$

where $q_{||}$ and q_z are the parallel and perpendicular distances from the nearest Bragg point in reciprocal space and η is the roughness exponent ($\eta \propto q_z^2$). The first term in

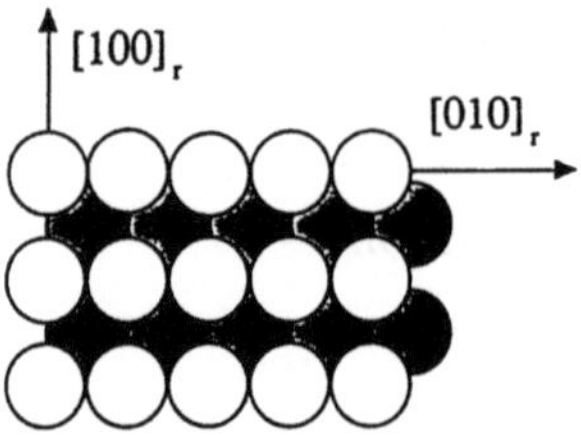

Fig. 1. Top view of the Ag(110) fcc surface. The rectangular coordinates system is shown (indicated by an index 'r'). Q is measured in units of [1.5377Å^{-1}, 2.1746Å^{-1}, 2.1746Å^{-1}]

eq. 2 is the crystal truncation rod (CTR) [8]. The second term is due to the roughness and broadens the CTR.

Held *et. al.* [3] studied Ag(110) in-plane lineshapes at $(1,0,Q_Z)_r$ for Q_Z close to zero (the index 'r' indicates rectangular coordinates, see Fig 1). They found satisfactory fits to eq.(2) over the whole temperature range. The Q_Z dependence, however, was not tested.

Experiments and Results. We looked at room temperature diffraction data at different points on the $(1,0)_r$ rod for perpendicular momentum transfer Q_Z=0.1, 0.4, 0.7. Parallel momentum scans were taken along the $[010]_r$ azimuth. Power-Law fits were only satisfactorily for Q_Z= 0.1, similar to Held *et. al.* [3]. At larger Q_Z, an obvious asymmetry develops that made the use of a two component lineshape necessary (see fig. 2). The χ^2 for the two component fits were smaller by at least a factor of 4.5.

Since both components increase in amplitude as Q_Z approaches the Bragg condition, local steps on a long range flat surface can be excluded. This is because steps give rise to a broad component that has a minimum at the Bragg point, where

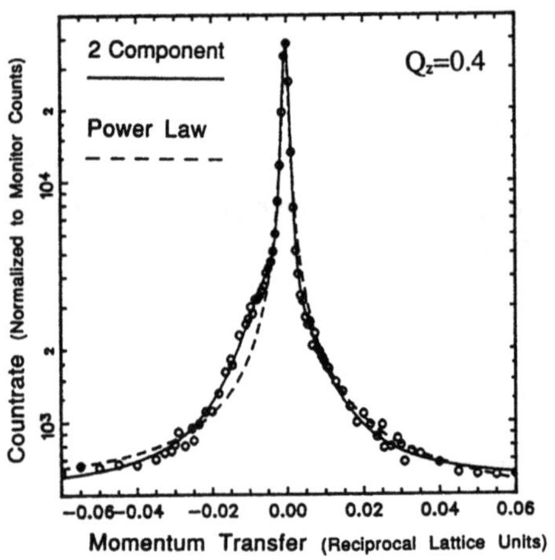

Fig. 2. Room temperature scans of the (10) beam in the $[010]_r$ direction at different Q_Z. The solid lines are the two component fits. Dashed lines are power law fits.

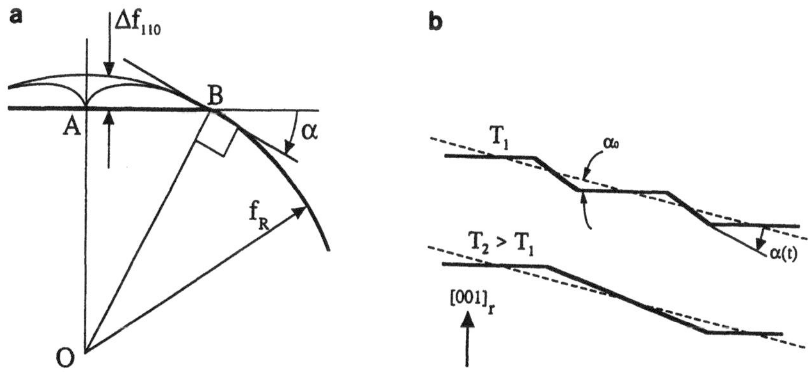

Fig. 3. (a) Polar plot of the free energy curve near the [110] direction. At temperatures above T_R the free energy surface is spherical. At lower temperatures a cusp develops in the [110] direction leading to the flat (110) phase. The thick line is the ECS. Orientations between A and B are forbidden. (b) Side view of the faceted surface at two temperatures. The rough inclined phase makes an angle α with the (110) plane that depends on the depth of the [110] cusp in the free energy curve.

all atoms scatter in phase [9]. The increase in intensity is instead consistent with two CTRs that have maxima at the Bragg point.

The theory of ECS can be used to understand this data [10,11]. As long as isolated steps can exist on the surface all crystal directions near (110) are present. If as in the case of these experiments on Ag(110), isolated steps congregate away from the facets, most crystal orientations close to (110) are forbidden. To see this let us assume that the free energy surface close to the melting point is spherical, and that a cusp develops in the [110] direction below the (110) roughening temperature (see fig. 3a) [10,12]. The ECS of such a system is the internal envelope of planes drawn perpendicular to the radius vectors at the intersection with the free energy plot [10]. Therefore, the cusp will lead to a flat region of (110) oriented surface. Note also, that all orientations in fig. 3a between A and B are forbidden since they do not contribute to the internal envelope. In this sense only points A and B are well defined. Any miscut leading to forbidden orientations must then phase separate into either the A (flat (110)) or the B (rough) phase ("hill-and-valley" structure), according to a lever rule [10,13]. Figure 3b shows a side view of the surface. In the diffraction profiles the sharp component arises from the flat (110) region while the broad component is due to an inclined region (see fig. 4). In this scenario the surface roughens by replacing the flat (110) terraces by the stepped facet. Note the symmetry at the in-plane position in fig. 4. This explains the symmetric lineshapes at Q_z=0.1. Further up the rod, the lineshape must be asymmetric and the intensity of the broad component increases, consistent with our data.

The angle between the flat (110) phase and the rough phase (tilt angle) can be evaluated from the separation of the two peak centers. To do this we took scans at $(1,0,0.5)_r$ as a function of $q_{||}$ in the $[010]_r$ direction, and fitted the data to a two component lineshape. The sharp component was fit by the square of a Lorentzian representing the convolution of the instrument function with the sample mosaic.

Fig. 4

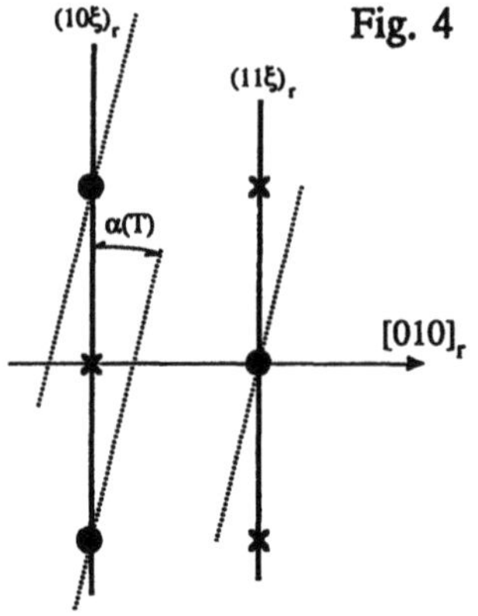

Fig. 5

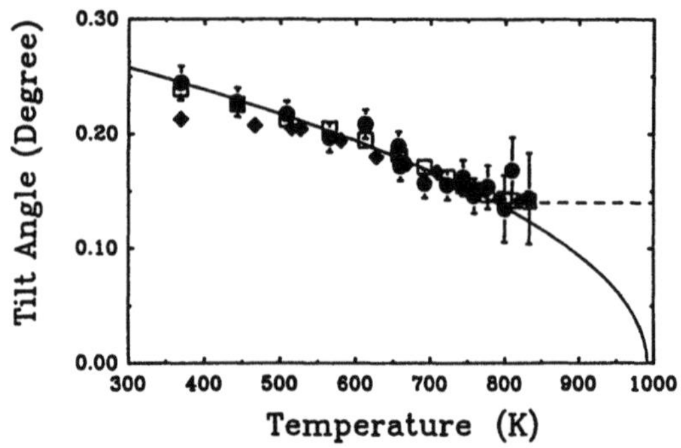

Fig. 4. Reciprocal space map showing the CTR's due to the two phases. The inclined dashed line corresponds to the stepped (rough) phase. The solid line is the CTR from the flat (110) phase. Circles and 'X' represent Bragg and Anti-Bragg points, respectively.

Fig. 5. Tilt angle α vs. temperature derived from the peak separation (filled circles) and by using the integrated intensities of the (1,0) beam (open squares) and (2,0) beam (filled diamonds). The solid line is a plot of eq. 3 with T_R=992 K.

The broad component is a single Lorentzian. Figure 5 shows a plot of the tilt angle versus temperature. At about 790K the tilt angle becomes a constant (α_o=0.14°), which agrees with the miscut of our sample determined by an optical measurement of the surface orientation.

From the geometry of fig. 3a it follows that for small α we have

$$\alpha(T) = \sqrt{\frac{2\,\Delta f_{110}(T)}{f_R}}. \tag{3}$$

If we assume a linear reduction of $\Delta f110$ (the depth of the cusp) with increasing temperature of the form,

$$\Delta f_{110} = |T_R - T|, \tag{4}$$

then the temperature dependence of α can be obtained after suitable adjustment of T_R and f_R (solid line in fig. 5). The roughening temperature of the (110) surface is found by extrapolating to the temperature where $\alpha(T)$=0. If we have a surface that has been miscut by an angle α_o and $\Delta f110$ is such that α_o is an allowed orientation, then only the B phase will exist. At these temperatures all crystal directions close the miscut are possible; isolated steps can persist and the surface is rough.

The relative amount of the two phases is also reflected in the integrated intensities of the two CTRs. Because the scattered intensity goes as the square of the number of scatterers, it is proportional to the surface area. By conserving the miscut of the sample, we have a second independent measure of the tilt angle

$$\alpha(T) = \alpha_0 \frac{(I_A+I_B)}{I_B} . \qquad (5)$$

These angles agree well with those derived from the peak separation, see fig. 5. The roughening transition is completed when the rough phase totally replaces the (110) flat phase. This temperature depends on the miscut of the sample.

Conclusions. The presented diffraction data can be satisfactorily fit to a two component lineshape. The results are interpreted as being due to two separate phases on the surface. A simple model has been developed that relates the tilt angle between the two phases with the roughening temperature T_R and the free energy f_R. The tilt angles calculated from the integrated intensities and the peak separations, respectively, were found to be in agreement with this model.

References

[1] J.C. Heyraud and J.J. Metois, Surf. Sci. **128**, 334 (1983); J.Cryst. Growth **50**, 571 (1990).

[2] S.T. Chui and J.D. Weeks, Phys. Rev. B **14**, 4978 (1976).

[3] G.A. Held, J.L. Jordan-Sweet, P.M. Horn, A. Mak, and R.J. Birgeneau, Phys. Rev. Lett. **59**, 2075 (1987).

[4] I.K. Robinson, E.H. Conrad and D.S. Reed, Journal de Physique (Paris) **51**, 103 (1990).

[5] J. Villain, D.R. Grempel and J. Lapujoulade, J. Phys. F **15**, 809 (1985).

[6] E.H. Conrad, L.R. Allen, D.L. Blanchard and T.Engel, Surf. Sci. **187**, 265 (1987).

[7] Y. Cao and E.H. Conrad, Phys. Rev. Lett. **64**, 447 (1990); H.-N. Yang, T.-M. Lu, and G.-C. Wang, Phys. Rev. Lett. **62**, 2148 (1989).

[8] I.K.Robinson, Phys. Rev. B **33**, 3830 (1986).

[9] C.S. Lent and P.I. Chohen, Surf. Sci. **139**, 121 (1984).

[10] M. Wortis, in *Chemistry and Physics of Solid Surfaces*, Vol. VII, ed. R. Vanselow and R. Howe (Springer, Berlin, 1988).

[11] G. Wullf, Z. Krist. 34 449 (1901); H. van Beijeren and I. Nolden, in *Structure and Dynamics of Surfaces II*, ed. W. Schommers and P. von Blankenhagen (Springer, Berlin, 1987).

[12] The assumed form of the cusp is such that it has a discontinuous derivative in the [110] direction. The surfaces roughens if the cusp disappears or rounds.

[13] B.M. Ocko and S.G.J. Mochrie, Phys. Rev. B **38**, 7378 (1988)
R.J. Phaneuf and E.D. Williams, Phys. Rev. Lett. **58**,2563 (1987)

Kinetics of Ordering with Random Impurities: Pb on Ni(001)

P.W. Stephens, P.J. Eng, and T. Tse

Department of Physics, State University of New York,
Stony Brook, NY 11974, USA

ABSTRACT: We apply high resolution x–ray diffraction to study the development of the c(2x2) phase of Pb on Ni(001). We observe lineshapes in agreement with the scaling theory of Ohta, Jasnow, and Kawasaki, but with an anomalously slow time dependence of average domain size R on time t, $R \propto t^{0.12}$. The presence of 1% – 3% carbon on the surface apparently acts as a weak random field.

Whereas equilibrium properties of two–dimensional phase transitions are generally well understood, less is known about the development of a system which is quenched into a nonequilibrium state. In this experiment, we study the time dependence of c(2x2) order of a monolayer of Pb on Ni(001). The adsorption geometry is illustrated in the inset to Fig. 1. Pb atoms adsorb onto one of the two equivalent sublattices. Long range order develops as Pb atoms jump between adsorption sites; consequently we speak of non–conserved order parameter (NCOP) dynamics. We study the late stages of NCOP domain growth, as the average domain size evolves from 100 to 1000 lattice spacings.

Our measurements were performed at the State University of New York X3B2 bending magnet x–ray beamline at the National Synchrotron Light Source (NSLS). A 6 mrad fan of 10.8 KeV x–rays were focused by two Si(111) crystals, the second one sagitally bent, to a spot 1.2 mm by 6 mm in the center of an ultrahigh vacuum chamber. The scattered

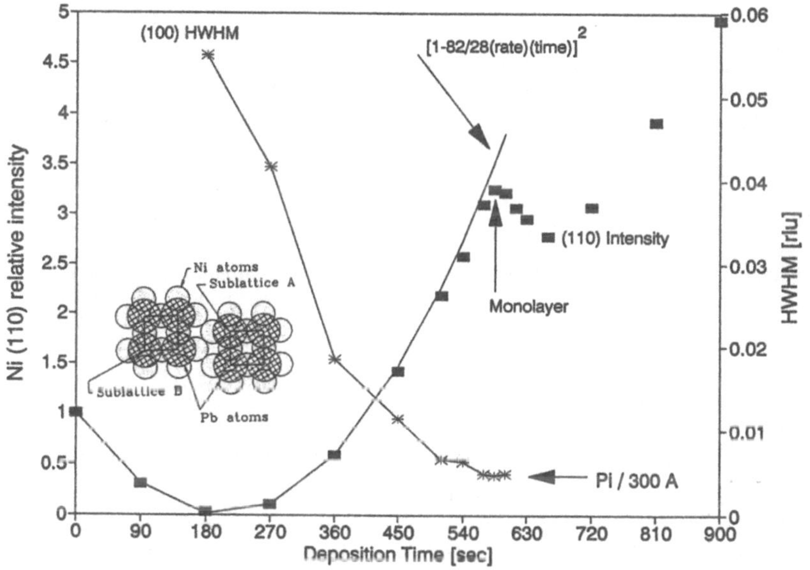

Fig. 1: Intensity of Ni (110) truncation rod (squares, left axis) and width of (100) superlattice ordering peak (asterisks, right axis) vs. Pb deposition time. Inset shows the adsorption geometry.

Eds.: H. Zabel and I.K. Robinson

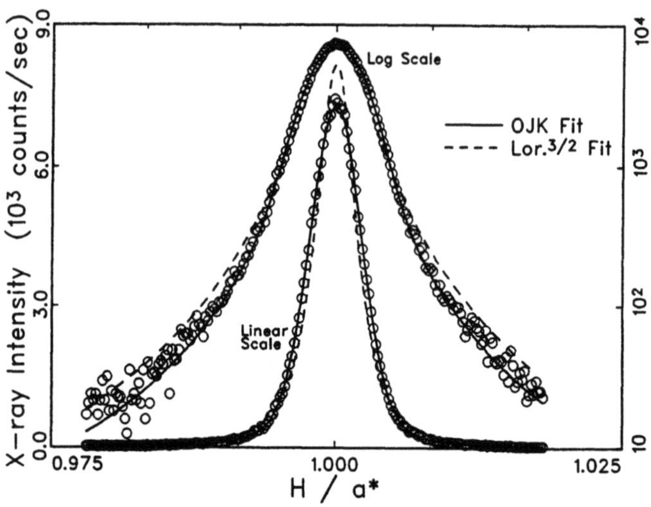

Fig. 2: Radial scan of (100) ordering peak for an up–quench to 200°C. The top curve is ploted using the right axis log scale; lower curve is ploted using the left axis linear scale. The solid line and broken line are fits the the Ohta lineshape and a Lorentzian to the 3/2 power, respectively, each convoluted with the measured resolution.

beam was collimated by a Ge(111) analyzer. The base pressure was typically 6 x 10^{-11} Torr, rising to 2 x 10^{-10} when the Pb deposition source was in use or the sample was being annealed. For the data reported here, the grazing angle of incidence α_i was approximately 0.3°, the critical angle for total external reflection. The Ni (001) substrate has a mosaic of 0.15° FWHM, and the surface is aligned to the average (001) direction to within 0.025°. The substrate was prepared by repeated cycles of Ar^+ sputtering and annealing to 700° C. The FWHM of the (11ℓ) truncation rod was 0.0013 $Å^{-1} = 2\pi$ / 4800 Å. Auger spectra showed that the Ni surface had no detectable impurities except for ~1% carbon at the time of Pb deposition. In contrast to previous LEED experiments on NCOP ordering in the presence of impurities,[1] the range of diffraction peak widths, 0.0166 – 0.0019 $Å^{-1}$ full width at half maximum (FWHM), is significantly wider than the instrumental resolution (0.00082 $Å^{-1}$ FWHM) or the limit from the substrate grain size (0.0013 $Å^{-1}$ FWHM).

We calibrate the initial dose by observing the interference between the Pb and the Ni (110) truncation rod, as shown in Fig. 1. This also allows us to determine that the Pb atoms adsorb into the (expected) fourfold hollow sites, insuring that we are observing two–sublattice ordering. At the deposition temperature, the Pb domain structure is frozen, lacking the thermal activation energy necessary to evolve further. We start the evolution of the system toward its long–range ordered equilibrium state by increasing (typically 6° C/sec) to a final temperature T_f in the range of 163 – 320° C, with an overshoot less than 2° C. For initial coverages $\rho < 1$, the width of the (110) peak is unchanged and intensity increases by less than 16%. This indicates that the Pb atoms are essentially all located in fourfold sites.

We have tried to fit the data with several model scaling functions, appropriately convoluted with the (measured) system resolution. For example, Fig. 2 compares the scaling form predicted by Ohta et al. for pure Ising NCOP ordering,[2] with a Lorentzian to the 3/2 power, as would result from exponential correlations in two dimensions. The Ohta et al. form always fits significantly better than any power of a Lorentzian.

Using the results of our lineshape fits, we plot the width Γ ($\propto R^{-1}$) of the resolution–corrected diffraction profiles in Fig. 3. It is clear that the data at all temperatures are accurately represented by a power–law growth form, $\Gamma(t) = A\,t^{-x}$, with exponent $x = 0.12 \pm 0.02$. This is vastly lower than the exponent $x = 1/2$ predicted and

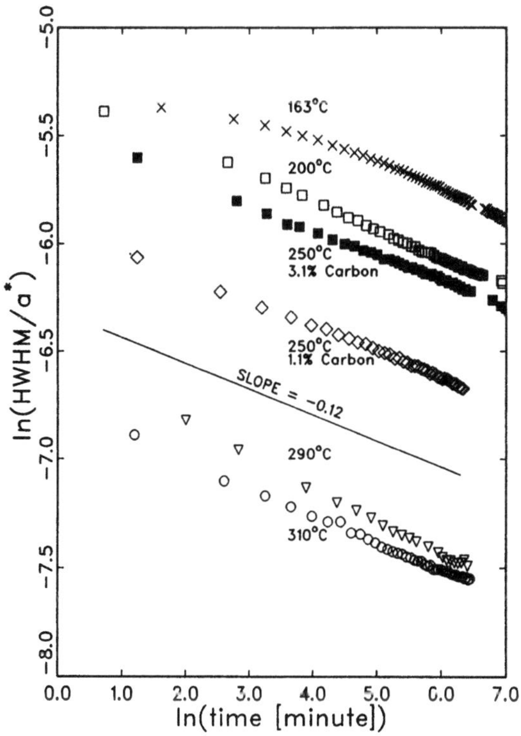

Fig. 3: The time dependence of the resolution–corrected (100) radial diffraction profile widths. Plotted are data for quenches to different temperatures with approximately 1.5% C impurity before Pb deposition.

observed for pure NCOP systems driven by domain–wall interface tension. This result indicates that a mechanism other than simple motion of adsorbed Pb atoms between the two equivalent sublattices is limiting the domain wall motion.

A priori, it appears that the origin of the difference between this experiment and the predictions of NCOP growth is the presence of a small amount of carbon impurity on the surface. Indeed, in order to check the influence of C on the ordering process, we have observed the evolution of the lineshape following an intentional increase in the C dose from 1% to 3%. The diffraction profiles continue to fit the OJK scaling form, and Γ follows the same power law in time (Fig. 3). Roughly tripling the amount of surface C changes only the growth rate A in the power–law evolution of $\Gamma(t)$, so that at a given time after t_0, the diffraction peak is a factor of 1.6 wider. We therefore infer that the unusual time dependence of Γ is a result of the dilute impurity C present on the surface.

NCOP growth in the presence of quenched impurities (random fields) has been considered by several workers; however, they have predicted either logarithmic corrections to the pure system $R = t^{1/2}$ growth law or else a much slower logarithmic $R = \log(t)$.[3] While a small power–law exponent is difficult to distinguish from a logarithm, we can exclude the latter case on the grounds of the independence of slope on temperature. Accordingly, we conclude that this is the first determination, either theoretical or experimental, of power–law growth (other than $t^{1/2}$) in an Ising NCOP system. Our interpretation of these results will be presented in a more detailed future publication.[4]

This research was supported by NSF grant DMR–8922066 and DOE grant DEFG0286ER45231.

REFERENCES

[1] J.–K. Zuo, G.–C. Wang, and T.–M. Lu, Phys. Rev. Lett. 60, 1053 (1988); Phys. Rev. B 40, 524 (1989).

[2] T. Ohta, D. Jasnow, and K. Kawasaki, Phys. Rev. Lett. 49, 1223 (1982).

[3] M. Grant and J.D. Gunton, Phys. Rev. B 29, 1521 (1984); ibid B 29, 6266 (1984); E.T. Gawlinski, S. Kumar, M. Grant, J.D. Gunton, and K. Kaski, ibid B 32, 1575 (1985).

[4] P.J. Eng, P.W. Stephens, and T. Tse, submitted for publication.

Part II

Reflectivity

Anomalous Reflectivity: A New Method for Determining Density Profiles of Thin Films

S.K. Sinha[1,], M.K. Sanyal[1], K.G. Huang[1], A. Gibaud[2], M. Rafailovich[3], J. Sokolov[3], X. Zhao[3], and W. Zhao[3]*

[1]Physics Department, Brookhaven National Laboratory, Upton, NY 11973, USA
[2]University of Maine, V.P. 807, C.N.R.S., Le Mans, France
[3]Physics Department, Queens College, NY 11367, USA
*Present address: Corporate Research, Exxon Research & Engineering Co., Annandale, NJ 08801, USA

Abstract. We present a method of obtaining a model-independent density profile for a thin film deposited on a substrate using reflectivity measurements carried out at energies close to, and away from, a convenient absorption edge of the substrate. We illustrate the method using simulated data from various model profiles and with a synchrotron X-ray experiment carried out on a polymer film deposited on a Germanium substrate.

1. Introduction

The use of X-ray and neutron reflectivity measurements to determine the density profile $\rho(z)$ across an interface has become increasingly popular over the last few years. Unfortunately, the fact that phase information is lacking and that one is therefore only measuring intensities necessitates the fitting of model profiles to the data. The most rigorous method is to divide the profile into appropriately fine slices of assumed constant scattering density normal to the interface and use standard optics methods [1] for calculating the reflectivity of a composite medium. Such a method becomes tedious and difficult to apply unless one has a good initial guess for the density profile. Alternatively, one can use the approximate formula, based on the Born Approximation [2]

$$R = R_F \left| \frac{1}{\rho_0} \int \frac{d\rho}{dz} e^{-iqz} dz \right|^2 . \quad (1)$$

In the absence of phase information, only symmetric density profiles can be uniquely obtained in this fashion.

The method we use is based on the fact that the reflectivity of a thin film on a substrate involves an interference between the reflectivity of the film and that of the substrate, and hence contains phase information. However, in order to obtain both the real and imaginary parts of $\rho(q_z)$, the Fourier transform of the (x,y) averaged scattering density, one needs also to make two different measurements between which one can vary the reflectance of the substrate in a known manner. This can be done with X-rays by measurements at, and away from, an absorption edge of the substrate. (It can also be accomplished for neutrons by changing the contrast, e.g., by deuteration, or by up- and down-spin reflectivity measurements from a magnetized substrate [3].)

Let us consider a zeroth order approximation $\rho_0(z)$ to the density profile, such that the (complex) reflection coefficient $r_0(k_1)$ for the system S_0 (substrate + $\rho_0(z)$) can be calculated <u>exactly</u>. (k_1 is the normal component of the incident wavevector.) In practice, $\rho_0(z)$ may usually be taken to be a constant equal to the average density of the film over the total thickness δ of the film, and $r_0(k_1)$ can then be calculated analytically. The actual density profile is $(\rho_0(z) + \Delta\rho(z))$ where $\Delta\rho(z)$ is now to be determined. (The film surface is taken as z = 0, with z increasing towards the substrate, which thus lies at $z = \delta$) Using the Distorted Wave Born Approximation (DWBA) [4,5], the reflectivity for the actual system is obtained as

$$R(k_1) = \left| \, i \, r_0(k_1) + \frac{4\pi b}{q_z} \left(a^2(k_1)\, \Delta\rho\, (\tilde{q}_z) + b^2(k_1)\, \Delta\rho^*(\tilde{q}_z)\right) + 2a(k_1)\, b\,(k_1)\, \Delta\rho\,(o) \right|^2 \tag{2}$$

where b is the scattering length of the electron (or nucleus), q_z (= $2k_1$) is the normal component of wavevector transfer in free space, and $\tilde{q}_z$ is the same quantity inside the medium, i.e., the uniform film of density ρ_0. $a(k_1)$ and $b(k_1)$ are the (complex) coefficients of the transmitted and reflected wave in the assumed uniform film and may be calculated by standard methods. If q_c corresponds to the critical wavevector for the uniform film of density ρ_0 (= $4b\rho_0$), then $\tilde{q}_z = (q_z^2 - q_c^2)^{1/2}$. $\Delta\rho(q)$ is (apart from a trivial factor) the Fourier transform of $\Delta\rho(z)$ given by

$$\Delta\rho\,(q) = \int_{-\infty}^{\infty} \Delta\rho\,(z)\, e^{iqz} dz\ . \tag{3}$$

Since $\Delta\rho(z)$ is real, $\Delta\rho(-q) = \Delta\rho^*(q)$, and so only the values of $\Delta\rho(q)$ for positive arguments are required to obtain $\Delta\rho(z)$. The term involving $\Delta\rho(o)$ vanishes identically if ρ_0 is chosen to be the true average density of the film and this can usually be done, at least iteratively. $r_0(k_1)$ can be changed by one of the methods discussed above. For X-rays, $r_0(k_1)$ will change at an X-ray absorption edge in a known manner, which is both measurable (in a subsidiary experiment involving the bare substrate) or calculable. From two measured reflectivity functions $R(k_1)$ at, and away from, the edge, one can then use Eq.(2) to solve for the real and imaginary parts of $\Delta\rho(\tilde{q}_z)$ for $\tilde{q}_z$ corresponding to each k_1. Finally, a Fast Fourier Transform is used to obtain the function $\Delta\rho(z)$, which when added to $\rho_0(z)$ yields the required real-space density profile $\rho(z)$ in a model-independent manner. (Note that measurements of reflectivities for uniformly spaced values of k_1 do not result in uniformly spaced values of $\tilde{q}_z$, but can be interpolated to lie on a uniform grid for the Fourier transformation).

2. Numerical Simulations

In order to test the method conceptually, we considered a germanium substrate with the roughness and refractive index parameters obtained in the present experiment (see below) and modelled adsorbed films of

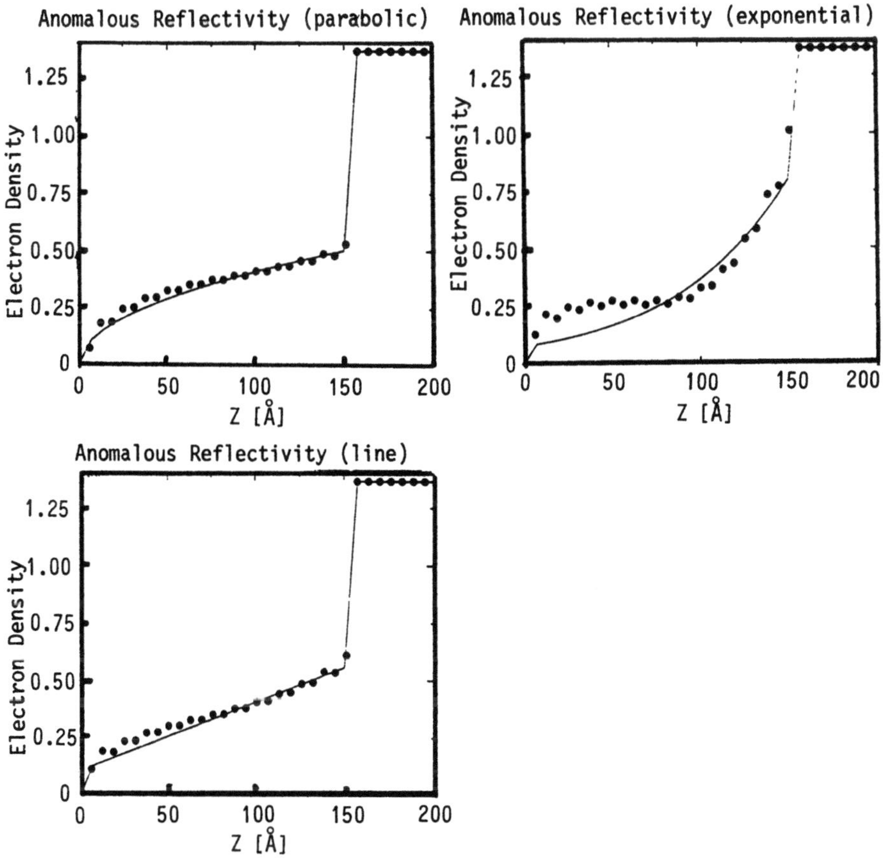

Figure 1 Theoretical model electron density profiles (solid curves) and profiles obtained by inverting simulated data using the present method (filled circles) for (a) parabolic, (b) exponential, and (c) linear density profiles. The units for electron density are electrons/$Å^3$.

various electron density profiles (choosing the order of magnitude of the electron density to be that of polystyrene) such as (a) an exponentially decreasing density profile, (b) a parabolic density profile, and (c) a linear density profile. We then simulated "measured" reflectivities by calculating these using the "slicing" method at the two different energies used in the actual experiments. Using as our zeroth-order approximation for $\rho_0(z)$ a uniform film of average density and thickness, we then used Eq.(2) to solve from these reflectivities for the density profiles as described above. Figure 1 shows that even one iteration yields a $\rho(z)$ which closely follows the actual model density profile, indicating the feasibility of the method. The deviations are due to the approximations inherent in the DWBA, as well as the limitations due to Fourier transforming data over a limited range of q. It should be noted that in these simulations, only simulated reflectivities up to a maximum value of $q_z = 0.5Å^{-1}$ were used, corresponding to actual experimental limitations. This restricts the spatial resolution for the density profile to be ~ (π/q_z (max)) or about 6Å for the present experiments. Obviously, spatial variations more rapid than this cannot be accessible from reflectivity measurements - a fact that is often ignored in conventional density profile "fits" to reflectivity data. The importance of using a highly polished substrate (so that $r_0(\vec{k}_1)$ in Eq.(2) is still significant at large values of q_z) for such measurements is to be noted.

3. Experimental Demonstration

We now discuss the application of the method to a real system, namely a film of 10% by volume monodisperse polybromostyrene (PBrS) of MW 20000 in solution in MW 4000 Polystyrene (PS) deposited on a polished Germanium substrate. The measurements were made at Beamline X22C at the National Synchrotron Light Source at Brookhaven. Reflectivity measurements were made at 11.030 keV close to the Germanium K-edge and at 10.600 keV. From a subsidiary measurement on the bare substrate at the same two energies, the roughness of the Ge surface (3.96Å) was determined, as well as the value of q_c for Ge at 11.030 keV (q_c = 0.0432Å^{-1}) and at 10.600 keV (q_c = 0.0472Å^{-1}), and the absorption coefficients at the two energies. The change of q_c corresponds to a decrease of 5.75 electrons for Ge at 11.030 keV, which is close to the maximum expected change of 8 electrons. (A 24.3Å oxide layer needed to fit the bare substrate data was apparently removed by the film deposition process since the reflectivity results could not be consistently analyzed by assuming that it was present.)

Figure 2 shows the total electron density profile obtained using our method with one iteration. A recalculation of the reflectivity curves at the two energies with this profile and the Ge substrate parameters, and using the slicing method gave a reasonably good fit to the observed reflectivity profiles, indicating that a solution had been obtained, as shown in Figure 3(a). Figure 3(b) shows the reflectivity profile at 10.6 keV using the "single uniform box" ($\rho_0(z)$) profile model for the film. From Figure 2 it is clear that there is a thin layer of surface segregated PBrS at the Ge surface, the electron density value for $\rho(z)$ there being consistent with that of pure PBrS. There is also a

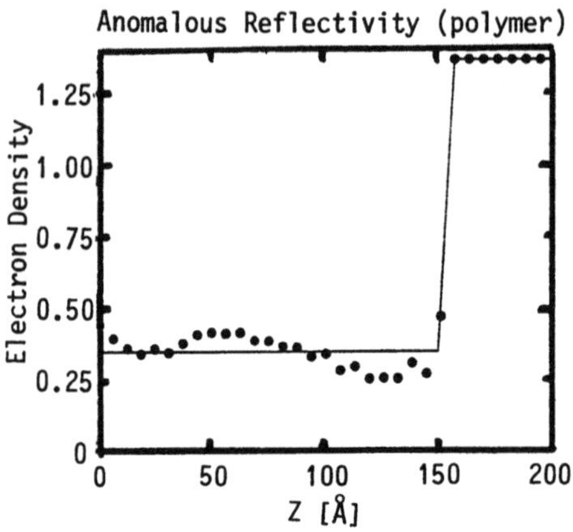

Figure 2 Electron density profile for polymer film adsorbed on Ge substrate (in electrons/A^3).

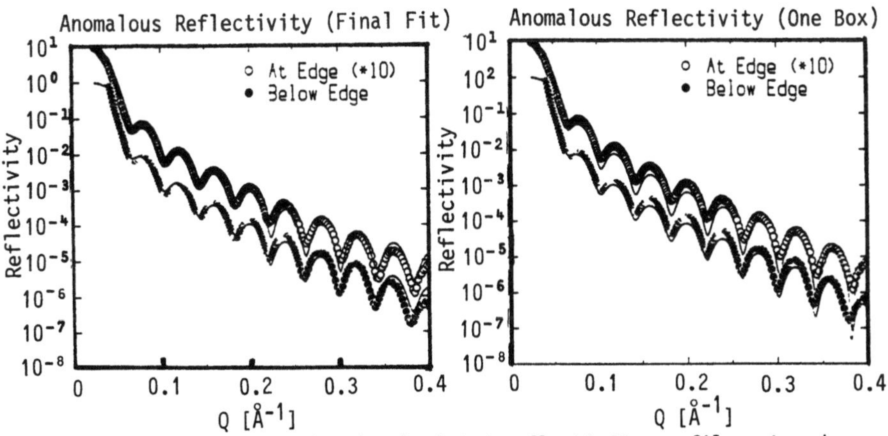

Figure 3 (a) Measured and calculated reflectivity profiles at and below the Ge K-edge (using slicing method) for polymer film with the electron density profile shown in Figure 2. The circles represent the experimental data and the solid curves the calculations.

(b) The same quantities calculated using a uniform (single-box) model for the film.

depleted region about 20Å thick beyond this layer for which we have no current explanation. (Model calculations showed that the data could not be fitted without this depletion layer.) Independent secondary ion mass spectrometry (SIMS) studies with 2 keV Ar^+ ions [6] also showed that there is a surface segregation of PBrS at the Ge surface, but the spatial resolution (~ 85Å) was not sufficient to show the depletion layer.

4. Conclusions

In this paper we hope to have demonstrated that the method of anomalous reflectivity does provide a unique model-independent method of obtaining the density profile of a film adsorbed on a substrate whose roughness is small and known and whose refractive index parameters are accurately known at energies close to, and away from, an edge or whose reflectivity can be varied by some other means. The method is similar in spirit to anomalous scattering techniques in X-ray crystallography, except that the profile is inverted in principle from continuous measured functions of wavevector transfer rather than at a set of discrete reflections.

This work was partially supported by US DOE DE-AC02-76CH00016.

5. References

[1] L. G. Parratt, Phys. Rev. 95, 359 (1954).
[2] J. Als-Nielsen, NATO ASI Ser. B176, 639 (1988); P. S. Pershan, General Discussion 89, Royal Society of Chemistry, Faraday Division (1990).

[3] G. P. Felcher, private communication; C. F. Majkrzak, private communication.
[4] A. Messiah, Quantum Mechanics (Wiley, New York, 1962), Vol. 2, p. 673.
[5] S. K. Sinha, E. B. Sirota, S. Garoff, and H. B. Stanley, Phys. Rev. B38, 2297 (1988).
[6] X. Zhao, W. Zhao, J. Sokolov, M. Rafailovich, S. A. Schwarz, and B. J. Wilkens, to be published.

Specular and Diffuse Scattering Studies of Multilayer Interfaces

M.K. Sanyal[1,a], *S.K. Sinha*[1,b], *A. Gibaud*[1,c], *S.K. Satija*[2], *C.F. Majkrzak*[2], *and H. Homa*[3]

[1]Department of Physics, Brookhaven National Laboratory, Upton, NY 11973, USA

[2]National Institute of Standards and Technology, MD 20899, USA

[3]Physics Department, Brooklyn College of CUNY, NY 11210, USA

Abstract. The results of specular and diffuse x-ray scattering studies of multilayers are discussed. We show here that such studies can yield detailed statistical information about the interfacial roughness and morphology. Results on a GaAs/AlAs multilayer are presented and the data is analyzed within the Born approximation.

1. Introduction

In this paper we present results on the characterization of interfacial roughness of multilayers using the x-ray reflectivity technique. Although detailed characterization of a multilayer system requires analysis of both reflectivity data at small angles and data in the (wide angle) Bragg reflection regime, we shall restrict ourselves here to discussing only the small angle region. This analysis gives us knowledge about the "global" interfacial roughness and its conformality over the entire multilayer. The analysis of the scattering data, which can be grouped into two categories namely specular and diffuse, is done within the Born approximation. In the small angle limit here we neglect the crystal structure of the materials assuming only uniform electron densities in the layers separated by (rough) interfaces.

2. The Model

It has been shown [1] that a knowledge of the height-height correlation function

$$C_0(R) = \langle \delta Z(0) \delta Z(R) \rangle \tag{1}$$

can be used to derive the specular and diffuse scattering intensities for solid as well as liquid [2] surfaces using the Born approximation. Here $\delta Z(R)$ [assumed to be a single valued function] is the height fluctuation of the surface above the plane at a lateral position R. In the case of multilayers we have to generalize this function taking into account the possibility of correlation between height fluctuations of different interfaces and we assume the form

$$C_{ij} \equiv < \delta Z_i(0) \delta Z_j(R) > = C_0(R) * \exp(-|Z_i - Z_j|/\xi_\perp) \ . \tag{2}$$

$\xi_\perp$ is the correlation length for the roughness perpendicular to the interface and

Z_i, Z_j are the mean positions of interfaces i and j respectively. The term $C_0(R)$ generally takes [1] the form $\sigma^2 \exp((-R/\xi_{\|})^{2h})$ with $(0 < h < 1)$. In the above expression we have assumed that the r.m.s value of the interfacial roughness is the same for all the interfaces. By using this height-height correlation function and also assuming some r.m.s random error δ in the deposition thickness of each layer one can derive the expression [3] for scattering intensity from N such interfaces as

$$I = \frac{I_0}{q_z^2} \frac{b^2}{K_0^2 \sin\alpha \sin\beta} \sum_{ij} \Delta\rho_i \Delta\rho_j \exp(-iq_z(Z_i - Z_j)) \\ * \exp[-q_z^2(\sigma^2 + \delta^2|i-j|)] \\ * \int\int dxdy \exp(q_z^2 C_{ij}(R)) e^{-iq_{\|}\cdot R}. \tag{3}$$

In the equation (3) $\Delta\rho_i$ represents difference between electron densities of the medium above and below the ith interface, α and β are respectively grazing angle of incidence and scattering, b is the Thomson scattering length given by e^2/mc^2 and K_0 is the wave vector of the incident radiation. It should be noted here that the above expression is not valid near the critical angle and has to be modified according to the distorted wave Born approximation to explain the scattering in that region. The integral in equation (3) can be split into two parts, one of which yields the specular reflectivity and the other yields the diffuse scattering where the integral in equation (3) is replaced by

$$\int\int dxdy[\exp(q_z^2\sigma^2 e^{-(R/\xi_{\|})^{2h}} e^{-|Z_i - Z_j|/\xi_{\perp}}) - 1]e^{-iq_{\|}\cdot R} . \tag{4}$$

3. Results and Discussions

We shall present here the data for a multilayer, which has 77 bilayers of GaAs and AlAs, to demonstrate the use of the above expressions. The experiment was performed at the X-22B beamline of NSLS at Brookhaven National Laboratory. While taking the data the resolution in the direction normal to the scattering plane was kept wide open. This effectively performed one integration and we thus have to deal with only a one dimensional integral to evaluate equation (4). In Fig. 1 we show the specular and longitudinal diffuse scattering (i.e. parallel to but displaced from the specular ridge) as a function of q_z. The presence of multilayer Bragg peaks in the diffuse scattering clearly demonstrates the conformality of the interfaces. It should be mentioned here that to obtain the "true" specular component of the scattering, one has to subtract the diffuse component from the measured specular data and correct it for the variation of the resolution function. Unless the above procedure is followed one can grossly underestimate the "global" roughness, since one is measuring only the "local" roughness determined by the instrumental resolution. The true specular component (away from the critical angle) and a fit with the curve calculated from equation (3) are shown in Fig. 2. The total bilayer thickness

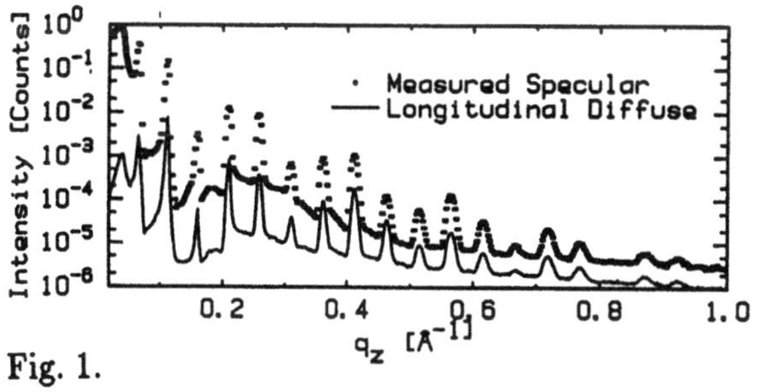

Fig. 1.
As measured specular reflectivity and longitudinal diffuse scattering data for GaAs/AlAs multilayer.

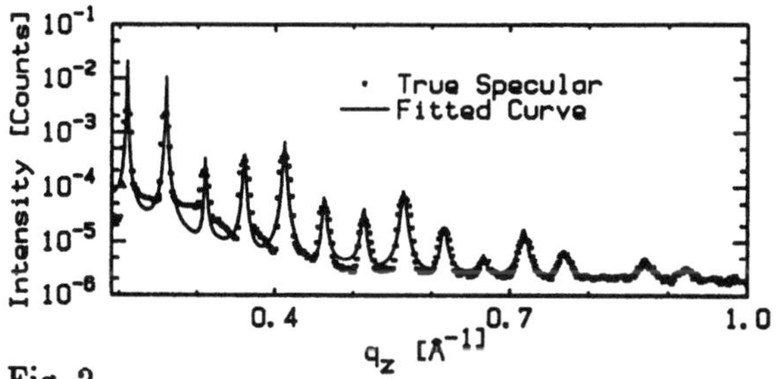

Fig. 2.
Corrected true specular reflectivity data and the fitted curve as described in the text.

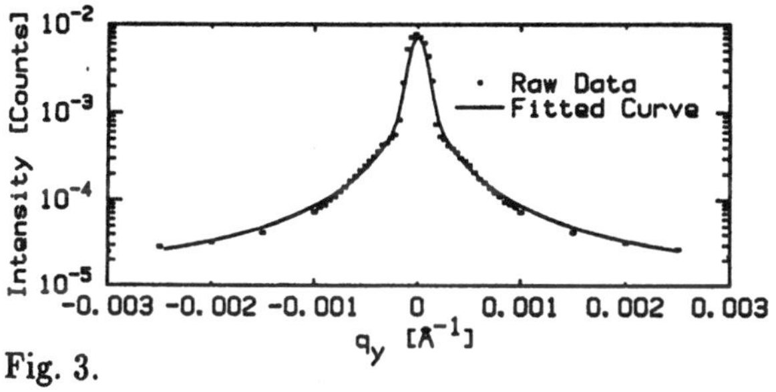

Fig. 3.
Transverse diffuse scattering data across the fifth Bragg peak. The solid curve represents the calculated scattering.

and the thickness ratio between the layers are obtained to be 122.80Å and 0.684 respectively; "global" roughness σ and cumulative roughness δ are found to be 1.6Å and 1.07Å respectively. In Fig. 3 we show analysis of a typical transverse diffuse scan. The specular component of the scan is assumed to be a Gaussian function with the width of the resolution function at that position. The diffuse component is calculated using equation (4) with a further assumption that correlation length $\xi_\perp$ is much larger than the multilayer thickness. The solid line in Fig. 3 shows the fitted curve generated by numerical integration of equation (4) and the obtained values of the parameters h and $\xi_{||}$ are found to be 0.4 and 7000 Å respectively. The small value of h, which contains information regarding the texture of the roughness, indicates that the interface is not a smooth one although it has a large in plane correlation length.

We have shown here that analysis of small angle specular reflectivity and diffuse scattering data of multilayer systems can generate information regarding the interfacial roughness, its conformality and texture. This in turn provides us valuable hints regarding the growth process for multilayers. Detail analysis of the specular and diffuse scattering data is beyond the scope of this short communication and will be presented in a future publication.

This work was supported by US DOE DE-AC02-76CH00016.

(a) On leave from : Solid State Physics Div., BARC, Bombay 400 085 India.
(b) Present address : Exxon Research and Engineering Co. NJ 08801, USA.
(c) Present address : Fac. Science, Univ. du Maine, 72017 LeMans, France.

4. References

1. S.K. Sinha, E.B. Sirota, S. Gasoff and H.B. Stanley, Phys. Rev. **B38**, 2297 (1988).
2. M.K. Sanyal, S.K. Sinha, K.G. Huang, and B.M. Ocko, Phys. Rev. Lett. 66, 628 (1991).
3. S.K. Sinha, M.K. Sanyal, A. Gibaud, S.K. Satija, C.F. Majkrzak, and H. Homa, Proceedings of NATO Advanced Study Institute Conference in Crete, Greece, June 26–July 6, 1990.

Scattering Cross-Section of X-Rays and Neutrons for Grazing Incidence onto Thin Films

A. Haase and *S. Dietrich*

Fachbereich Physik, Bergische Universität Wuppertal,
Postfach 100127, W-5600 Wuppertal 1, Fed. Rep. of Germany

Abstract. Within the Distorted-Wave-Born-Approximation the kinematic scattering cross-section of x-rays and neutrons impinging on a thin film at grazing incidence is expressed in terms of the Fourier- and Laplacelike-transformed two-point correlation function of the number density in the film. This expression takes into account the multiple scattering processes which arise due to the difference between the indices of refraction of the film and those of the two adjacent media. *Inter alia*, the scattering data allow one to infer information about the lateral structures of the two confining interfaces of the film.

1. Introduction

Structural properties of condensed matter vary significantly as function of the spatial dimension of the system. Thin films offer the opportunity to study the smooth crossover between two and three dimensions upon increasing the film thickness. Such films either occur as a consequence of thermal equilibrium in the presence of suitable boundary conditions or they are manufactured, e.g., by expitaxial growth far from thermal equilibrium. In the first case they naturally arise close to wetting transitions, at which the $\alpha - \gamma$ interface between two phases α and γ is split into $\alpha - \beta$ and $\beta - \gamma$ interfaces by the intrusion of a β–like film of a third phase β [1]. As an example, α and β may be the vapor and liquid phase, respectively, of a one-component fluid and γ is an inert substrate as a spectator phase. In the case of surface melting α, β, and γ correspond to the vapor, liquid and solid phase, respectively, of one material. There is a large variety of examples ranging from the wetting of a substrate by various other solid phases to binary liquid mixtures, grain boundaries, liquid crystals, polymers etc.. *Inter alia*, upon depinning the emerging $\alpha - \beta$ interface is expected to display long-range lateral correlations [1], which so far have not yet been probed experimentally. On the other hand the aforementioned films, which are produced artificially, are of great technological importance like, *e.g.*, thin diamond, semiconductor, and magnetic films.

By using either light, x-rays or neutrons the mean density profiles within such films have been studied intensively in recent years by analysing reflectivity data.

Springer Proceedings in Physics, Vol. 61 Surface X-Ray and Neutron Scattering
Eds.: H. Zabel and I.K. Robinson

However, this gives no information about the interesting lateral structures which cause diffuse scattering around the specular beam. Such experiments have been already carried out successfully concerning the lateral structure of a free surface close to a second-order phase transition in its semi-infinite bulk [2]. Therefore the prospects for similar studies of films are encouraging. In order to provide a sound basis for interpreting such scattering data we have extended the Distorted-Wave-Born-Approximation for a semi-infinite geometry [3] to films.

2. The Scattering Cross-Section

With the notation $(\alpha, \beta, \gamma) \equiv (1,2,3)$ we consider the situation that the phases 1 and 3 occupy the half-spaces $V_1 = \{\mathbf{R} = (\mathbf{r}, z) \in \mathbb{R}^3 \mid z < 0\}$ and $V_3 = \{\mathbf{R} \in \mathbb{R}^3 \mid z > d\}$, respectively, whereas the β–like phase forms a film $V_2 = \{\mathbf{R} \in \mathbb{R}^3 \mid 0 < z < d\}$ of thickness d. Neutrons of mass m and with wave vector $\mathbf{K}_i \equiv \mathbf{K}_1^i = (\mathbf{k}_i, K_1 \sin \alpha_i)$ penetrate through V_1, which is taken to be a homogeneous and non-fluctuating medium; α_i is the angle between $\mathbf{K}_i$ and the plane $z = 0$. In $V_+ = V_2 \cup V_3$ the neutrons are exposed to a potential energy $U(\mathbf{R}) = b\, 2\pi(h^2/m)\rho(\mathbf{R})$ where $\rho(\mathbf{R})$ is the number density of the nuclei and b their scattering length; $\rho(\mathbf{R}) = \bar{\rho}(z) + \tilde{\rho}(\mathbf{R})$ where $\bar{\rho}(z)$ is the mean number density profile which is approximated by constant values ρ_1 for $z < 0$, ρ_2 for $0 < z < d$, and ρ_3 for $z > d$. Extending the analysis in Ref. 3 one obtains for the total scattering cross-section at nonzero parallel momentum transfer $\mathbf{p} = \mathbf{k}_i - \mathbf{k}_f$, where $\mathbf{K}_f \equiv \mathbf{K}_1^f = (\mathbf{k}_f, -K_1 \sin \alpha_f)$, $K_f = K_i$, gives the direction of the solid angle $d\Omega$,

$$\frac{d\sigma}{d\Omega} = b^2 A \int_0^\infty dz \int_0^\infty dz' f(z) f^*(z') G(\mathbf{p},z,z'). \tag{1}$$

A is the area of the film, $G(\mathbf{p},z,z') = \int d^2r\, e^{i\mathbf{p}\mathbf{r}} G(\mathbf{r},z,z')$ with

$$G(\mathbf{r}-\mathbf{r}',z,z') = \langle \tilde{\rho}(\mathbf{R}) \tilde{\rho}(\mathbf{R}') \rangle \tag{2}$$

denotes the thermally averaged two-point correlation function, and

$$f(z) = \Theta(d-z)[S_f e^{iq_{2,f} z} + S_f' e^{-iq_{2,f} z}]\,[S_i e^{iq_{2,i} z} + S_i' e^{-iq_{2,i} z}] + \Theta(z-d) T_i T_f e^{i(q_{3,i}+q_{3,f})z}. \tag{3}$$

Here $q_{\eta,i(f)} = iK_1[\sin^2\alpha_{c,\eta} - \sin^2\alpha_{i(f)}]^{1/2}$, $\eta = 2,3$, $q_{1,i(f)} = K_1 \sin \alpha_{i(f)}$, and $\alpha_{c,\eta}$ are the angles of total reflection for semi-infinite media 2 and 3, respectively. We assume that α_i or α_f is smaller than $\alpha_{c,3}$. Θ is the Heaviside function. In Eq. (3) $T = 4\,\Lambda q_1 q_2\, e^{i(q_2-q_3)d}$, $S = 2\,\Lambda q_1(q_2+q_3)$, $S' = 2\,\Lambda q_1(q_2-q_3)\, e^{2iq_2 d}$, and $\Lambda^{-1} =$

$(q_1+q_2)\cdot(q_2+q_3) + e^{2iq_2d}(q_1-q_2)(q_2-q_3)$ have to be evaluated at α_i and α_f according to the indices i and f, respectively. T is the transmission coefficient for the density profile $\bar{\rho}(z)$ whereas S and S' are the amplitudes of the wave functions generated by multiple scattering at the $\alpha-\beta$ and $\beta-\gamma$ interfaces. For $\rho_2=\rho_3$ or $\rho_1=\rho_2$ the expressions reduce to the known result for the semi-infinite geometry [3]. The discussion of various special cases, the formulae for x-ray scattering, the inclusion of absorption effects and the case that also V_1 exhibits fluctuations is given in Ref. 4.

References

1. S. Dietrich, in *Phase Transitions and Critical Phenomena,* edited by C. Domb and J. Lebowitz (Academic, London, 1988), Vol 12, p. 1.
2. L. Mailänder, H. Dosch, J. Peisl and R.L. Johnson, Phys. Rev. Lett. **64**, 2527 (1990).
3. S. Dietrich and H. Wagner, Z. Phys. **B56**, 207 (1984).
4. A. Haase and S. Dietrich, unpublished; A. Haase, Diploma Thesis, University of Wuppertal (1991).

Total Neutron Reflection: Experiments and Analysis

G.P. Felcher, W.D. Dozier, Y.Y. Huang, and X.L. Zhou

Argonne National Laboratory, Argonne, IL 60439, USA

The current status of the analysis of neutron reflectivity data is presented. The advantage of approximating the chemical depth profile by linear segments (rather than a histogram) is illustrated. An alternative method for obtaining the reflectivity analytically is derived and its performance compared to that of the Distorted Wave Born Approximation. Methods for inverting the measured reflectivity to obtain the profile are also discussed.

1. Activity at the Intense Pulsed Neutron Source

At Argonne National Laboratory, a complex of two reflectometers is operating at the Intense Pulsed Neutron Source. The complex (Fig.1) consists of a polarized neutron unit (POSY I) and a highly automated unit for unpolarized neutrons (POSY II). Their detailed description has been already presented in the literature [1][2], as well as the documentation of the research carried out with them. In the past year, most of the research of POSY I has been on magnetic multilayers [3][4], while POSY II (which operates most of the time in the user mode) was used to determine the equilibrium configurations and the diffusion properties of polymer blends [5][6][7].

In all these endeavours a significant amount of time was spent in converting the data into depth profiles of the relative density (chemical or magnetic) of different species. It is an understatement to say that the data analysis is not straightforward. The measured reflectivities

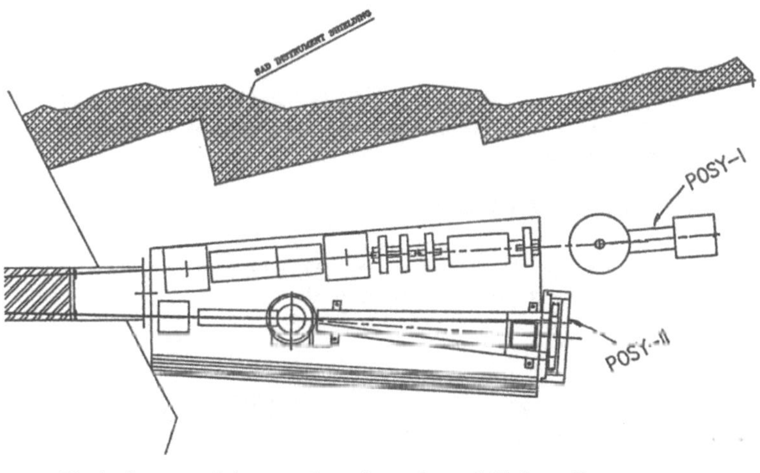

Fig.1. Layout of the complex of two time-of-flight reflectometers at IPNS (top view). In the upper unit, neutrons are polarized by reflection in a Co-Ti supermirror. The sample is temperature controlled (13-300 K) and in a variable magnetic field (0-8 kOe). Polarization analysis is included. The lower unit has computer-controlled collimation and an automatic sample changer. Both detectors are of the position-sensitive type.

are compared with those of model profiles, with features restricted by the findings of other techniques. However, even the fitting of a few parameters is quite laborious, because the reflectivity is not a linear transform of depth-dependent potential V(z) - at least for small values of the momentum transfer. To stimulate progress on such problems, a workshop was held at Argonne in 1990 [2], and here we present some of the ideas it inspired.

2. Approximating the depth profile

To calculate the reflectivity [8] of a profile V(z) it is convenient to divide the depth field into n intervals, such that within each interval the potential can be substituted by a function $V_n(z)$ for which the wavefunction can be expressed in an explicit analytical form. The most common method used is to divide V(z) into a histogram; within each segment n of z the wavefunction takes the form

$$\psi = a \exp(ik_n z) + b \exp(-ik_n z) \qquad (1)$$

where

$$k_n = (k_o^2 - 4\pi V_{n)})^{1/2} \qquad (2)$$

and the coefficients a and b are determined by the conditions at the boundaries, where the wavefunctions and their derivatives are supposed to be continuous. The reflectance r of a wave coming from the left is given by

$$\psi = \exp(ik_o z) + r \exp(-ik_o z) \quad . \qquad (3)$$

In Eq.(2), k_o indicates the component of the neutron's momentum perpendicular to the surface. The reflectivity $|r|^2$ is practically obtained solving for all a, b's of the system either with the matrix method or by Parratt's iterative formula [8].

It has been pointed [8] out for that several other functions V(z) the wavefunction ψ has an analytical form. This is the case when V(z) varies linearly, with slope $\Delta V/\Delta z$, between two values z_1 and z_2, respectively with ordinates V_1 and V_2. The solution of the Schroedinger equation is, in this case,

$$\psi = C_1 \,\mathrm{Ai}(y) + C_2 \,\mathrm{Bi}(y) \qquad (4)$$

where Ai, Bi are the Airy functions and the constants C_1 and C_2 are determined by the boundary conditions. The Airy functions are expressed in terms of a variable y, whose explicit functional dependence from the potential is

$$y = -(\Delta z/\Delta V)^{2/3} \left(k_o^2 - 4\pi(V_1+\Delta V/\Delta z)\right) \quad . \qquad (5)$$

The advantage of using the Airy functions, rather than an histogram is illustrated by an example. Suppose that the profile is given by an exponential:

$$V(z) = V_1 \exp(- z/z_0) + V_2 \qquad (6)$$

where, for instance, $V_1 + V_2 = 6 \times 10^{-6}$ Å^{-2}, and $V_1 = 1.5 \times 10^{-6}$ Å^{-2}; these values resemble rather closely those for deuterated and protonated polymers in neutron reflection [6]. Fig.2a shows the profile for a decay length z_o= 100Å, and Fig.2b the corresponding reflectivity. Also in the figures are presented the results of approximating the exponential function. The total height of V(z) has been cut in three equal parts, and in each of them the function has been substituted either with a step function or with a segment. The reflectivity obtained in the latter approximation is already quite close to the correct reflectivity (except quite close to the critical value of k_o). In other words, a model formed of just a few segments can realistically describe the depth profile.

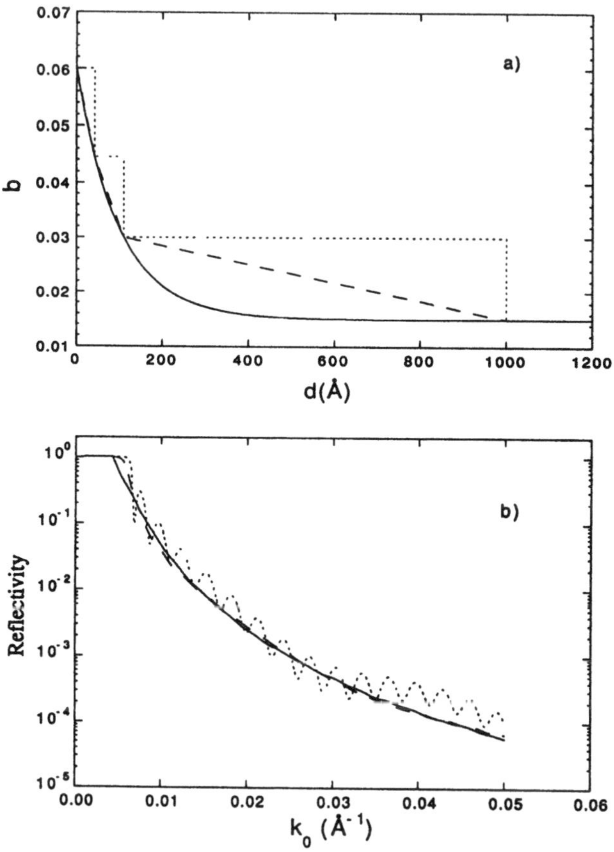

Fig.2 A chemical profile varying exponentially with the distance from the surface (full line, Fig.2a) and its reflectivity (Fig.2b) The dotted and dashed curves are respectively a histogram of the profile, and its segmented approximation.

3. The direct problem

For an arbitrary profile the reflectivity may not have a closed analytical form. However it has been found useful to derive approximate analytical solutions to the wave equation, such as for instance that provided by the first Born approximation. A new formulation has been developed recently [9], which gives a better approximation for values of k_0 close to the critical region. The starting point is to write the reflectance of the sample layer, of thickness d on a known substrate, in integral form:

$$r = R + \frac{i}{2k_0}\int_{-d}^{0} f(z)U(z)[e^{ik_0 z} + e^{-ik_0 z}] \tag{7}$$

where R is the reflectance of the bare substrate, $f(z)=-4\pi V(z)$ and U(z) is the neutron's wavefunction. This in general can be written as

$$U(z) = A(z)\exp\left(\int_0^z i\kappa_+ dz\right) + B(z)\exp\left(-\int_0^z i\kappa_- dz\right) \tag{8}$$

provided that A obeys the equation

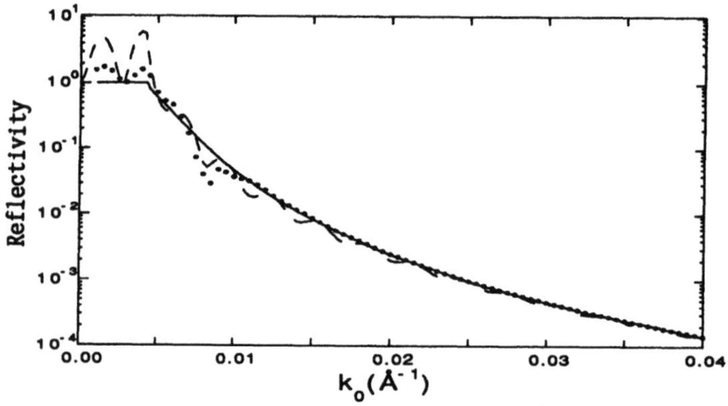

Fig. 3. The reflectivity calculated for the depth profile of Fig. 2a. Full line: exact calculation; dashed line: DWBA approximation; dots: the new approximation explained in the text.

$$A'' + 2ikA' + ik'A = 0 \tag{9}$$

and B obeys a similar equation. After defining a new differential operator $L_a = d/dz$, Eq.(9) is substituted by the more restrictive equation

$$L_a^2 + 2ik\, L_a + i\, k' = 0 \ . \tag{10}$$

The wavefunction now becomes

$$U(z) = A_0\left[\exp\left(\int_0^z i\kappa_+ dz\right) + r_b \exp\left(-\int_0^z i\kappa_- dz\right)\right] \tag{11}$$

with

$$\kappa_\pm = k\sqrt{1 \pm ik'/k^2} \tag{12}$$

$$A_0 = \frac{t_f \exp(-ik_0 d) \times \exp\int_{-d}^{0} i\kappa_+ dz}{1 + r_b r_f \exp\int_{-d}^{0} i(\kappa_+ + \kappa_-)dz} \quad ; \tag{13}$$

r_f and t_f are the effective Fresnel reflectance and transmittance at the front interface and r_b the Fresnel reflectance at the back interface, as defined by

$$r_f = \frac{k_0 - \kappa_-}{k_0 + \kappa_+} \quad \text{and} \quad t_f = \frac{2k_0}{k_0 + \kappa_+} \qquad \text{at } z = -d \ , \tag{14a}$$

$$r_b = \frac{\kappa_+ - k_s}{\kappa_- + k_s} \qquad \text{at } z = 0 \ . \tag{14b}$$

By passing from Eq.(9) to Eq.(10) it was implicitly assumed that $L_a'=0$ (as well as $L_b'=0$). In spite of this constraint, the new approximation is quite powerful. It contains as a special case the Distorted Wave Born Approximation, which is obtained simply by substituting in Eqs.(11)-(14) $\kappa_+=\kappa_-$ with an appropriate constant. The goodness of the new approximation is tested by calculating the profile of Fig.2a, and comparing it with that obtained by the exact solution as well as the DWBA (Fig.3)

4. The inverse problem and the phase

The object of the reflectivity measurements is to determine the depth profile, and this most ideally can be achieved by a mathematical procedure that would invert the data uniquely and accurately. Formally that can be obtained by recasting Eq.(7); the quantity that is sought is f(z) as a function of $r(k_0)$. There are well-developed methods for the inverse problems for quantum scattering from specific forms of potential field. However, in calculating the profile of a structured film deposited on a substrate, the high non-linearity in the total reflection region makes it difficult to derive a simple inverse operator. At Argonne an attempt is being made [9] in developing an algorithm based on the iterative solution of the straightforward transform of Eq.(7). In parallel with these efforts, the general approach of Marchenko and Gel'fand-Levitan [10] has been applied elsewhere [11] with extremely promising results.

The inversion methods tested up to now assume the full knowledge of the reflectance $r(k_0)$, both in amplitude and phase. The experimental quantity more readily obtained in a reflectivity experiment is $|r|^2$. It has been shown that if only this quantity is known, ambiguitities arise in the determination of the profile. While it is not clear at the present moment if the phase is required for all the reflectance spectrum, or only part of it, several attempts are afoot to devise experiments capable of providing phase information. In a companion paper, M. J. Bedzyk shows the capability of the standing wave technique. In principle, neutron spin precession can be used to probe the relative phases of layers reflecting at different depths [12]. Clever use can be made of the substrate as a reference reflector with "tunable" amplitude. In these proceedings, S. K. Sinha shows how to utilize the variation of the X-rays reflecting properties of the substrate over a resonance region. The neutron analog is a ferromagnetic substrate, with a reflectivity dependent on the polarization state of the neutron spin [13].

Acknowledgements This work is supported by the U.S. Department of Energy, Basic Energy Sciences, Materials Sciences, under contract No. W-31-109-ENG-38.

References

[1] G. P. Felcher, R. O. Hilleke, R. K. Crawford, J. Haumann, R. Kleb and G. Ostrowski, Rev. Sci. Instr. **58**, 609 (1987)

[2] Workshop on Methods of Analysis and Interpretation of Neutron Reflectivity Data, Argonne Aug. 25, 1990. Proceedings in:Physica **B173** (August 1991)

[3] S. S. P. Parkin, A. Mansour and G. P. Felcher, Appl. Phys. Lett. **58**, 1473 (1991)

[4] Y. Y. Huang, G. P. Felcher and S. S. P. Parkin, J. M. M. M. **99** (Sept. 1991)

[5] T. P. Russell, S. H. Anastasiadis, A. Menelle,G. P. Felcher and S. K. Satija, Macromolecules **24**, 1575 (1991)

[6] A. Karim, B. H. Arendt, G. P. Felcher and T. P. Russell, Thin Solid Films **202** (June 1991)

[7] G. P. Felcher, A. Karim and T. P. Russell, Journ. Non-Crystal. Sol.**131**, 703 (1991)

[8] J. Lekner, Theory of Reflection, Martinus Nijhoff Publ., Dordrecht 1987

[9] X. L. Zhou, S. H. Chen and G. P. Felcher, NATO Res. Workshop on Inverse Problems in Scattering and Imaging, Cape Cod April 14, 1991

[10] K. Chadan and P. C. Sabatier, Inverse Problems in Quantum Scattering Theory, Springer Verlag, N. Y. 1977

[11] P. Sacks, private communication. His computational method has been suggested by Bayliss, Li and Morawetz, Math. Comp. **52**, 321 (1989)

[12] C. F. Majkrzak, in ref.[2]

[13] W. D. Dozier, J. M. Carpenter and G. P. Felcher, Bull. Am. Phys. Soc. **36**, 772 (1991)

Profile Refinement in Neutron Reflectivity and Grazing Angle Diffraction

J.F. Ankner

National Institute of Standards and Technology,
Materials Science and Engineering Laboratory,
Building 235 E151, Gaithersburg, MD 20899, USA

Abstract. Using modern high-speed computers, one can now model specular reflectivity experiments with unprecedented precision. For thin film and multilayer systems exhibiting large characteristic interfacial roughness or interdiffusion and for polymeric materials with unusual density profiles, a discrete layer-modeling scheme is required for an accurate and physically meaningful interpretation of the modeled profile. The incorporation of such a layer-generating algorithm into a non-linear least squares fitting routine can greatly increase the quality of the information available in reflectivity measurements. Applying these techniques to the modeling of grazing-angle diffraction data is particularly important for neutrons since the characteristic penetration depths are, in general, some four times larger than for x rays. We will outline the grazing-angle analogue of reflectivity profile analysis using the distorted-wave approximation and discuss possible applications.

1. Introduction

X-ray and neutron reflectometry have become increasingly important techniques for the study of surface and interfacial phenomena [1–3]. As instruments improve and as more complex systems come under scrutiny, the ability to model the data becomes crucial to further development of the field. In particular, for studying systems with a large amount of interdiffusion or those exhibiting unusual scattering density profiles (such as polymers and magnetic films), one must employ schemes more sophisticated than the conventional Debye-Waller-type formalism [4]. Happily, digital computing power has greatly increased during the past decade and the approximations once deemed necessary in analyzing data are in many cases no longer required. Using currently available minicomputers (e.g. DEC 5800 series) and workstations (e.g. HP 700 series), one can now construct explicit and detailed interfacial models and incorporate these models within least-squares fitting routines. In sections 2 and 3, I will first outline the generic specular reflectivity boundary value problem and then discuss specific issues related to the practical implementation of modeling programs.

The use of grazing-angle diffraction has largely been limited to x-ray studies of surface phenomena, but an approach analogous to that used for reflectivity can be used to model subsurface in-plane structure as well. The scattering of subsurface layers is especially important for neutrons because of the intrinsically larger penetration depths of neutrons in matter. In section 4, I will present a depth-resolved derivation of the grazing-angle diffracted intensity within the distorted-wave approximation and discuss how through this formalism one might apply the methods of specular reflectivity profile refinement to the study of subsurface in-plane structure.

2. Boundary Value Problem

The specular reflectivity of neutrons or x-rays from a refractive boundary is described by the one-dimensional wave equation. The solution of this problem for multiple layers is conventionally given by either optical transfer matrices [5] or a recursion relation [6]. The recursion relation is the more compact of these two equivalent approaches and I will therefore make use of it in the following discussion.

A neutron plane wave $\vec{k}_0$ incident onto the surface of a material with a depth-dependent refractive index will excite both reflected and refracted beams within that material [7]. At the interface between two different media, the neutron wave function is [see Fig. 1(a)]

$$\Psi(\vec{r}) = \begin{cases} t_{l-1}e^{i\vec{k}_{l-1}\cdot\vec{r}} + r_{l-1}e^{i\vec{k}_{sl-1}\cdot\vec{r}} & (l-1)^{st} \text{ layer} \\ t_l e^{i\vec{k}_l\cdot\vec{r}} + r_l e^{i\vec{k}_{sl}\cdot\vec{r}} & l^{th} \text{ layer,} \end{cases} \tag{1}$$

where $\vec{k}_l$ and $\vec{k}_{sl}$ are the transmitted and specularly reflected wavevectors in refracting layer l and t_l and r_l their amplitudes. According to Snell's Law, only the wavevector components normal to the surface change in specular scattering,

$$\vec{k}_l = \vec{k}_{\|0} + \vec{k}_{\perp l} = \vec{k}_{\|0} + \left(\sqrt{k_{\perp 0}^2 - 4\pi N_l b_l}\right)\hat{n}_\perp, \tag{2}$$

where N_l is the number density of scatterers and b_l their nuclear scattering length, in general a complex quantity. The subscripts refer to wavevector components parallel and perpendicular to the sample surface. The critical angle for total external reflection is related to the real part of the scattering density $N_l b_l$: $\phi_{cl} = \sin^{-1}\lambda[\mathrm{Re}(N_l b_l)/\pi]^{1/2}$. One obtains the amplitudes t_l and r_l by applying as boundary conditions the continuity of Ψ and $\vec{\nabla}\Psi\cdot\hat{n}_\perp$ and recalling that $\vec{k}_{sl} = \vec{k}_l - 2(\vec{k}_l\cdot\hat{n}_\perp)\hat{n}_\perp$ [6],

$$t_{l-1}e^{ik_{\perp l-1}d_{l-1}/2} + r_{l-1}e^{-ik_{\perp l-1}d_{l-1}/2} = t_l e^{-ik_{\perp l}d_l/2} + r_l e^{ik_{\perp l}d_l/2}, \tag{3}$$

$$(t_{l-1}e^{ik_{\perp l-1}d_{l-1}/2} - r_{l-1}e^{-ik_{\perp l-1}d_{l-1}/2})k_{\perp l-1} = (t_l e^{-ik_{\perp l}d_l/2} - r_l e^{ik_{\perp l}d_l/2})k_{\perp l}, \tag{4}$$

where d_l is the thickness of refracting layer l. Solving the above equations for the ratios r_{l-1}/t_{l-1} and r_l/t_l one obtains the well-known Parratt recursion formula,

$$R_{l-1} = a_{l-1}^4\left[\frac{R_l + F_{l-1,l}}{R_l F_{l-1,l} + 1}\right], \tag{5}$$

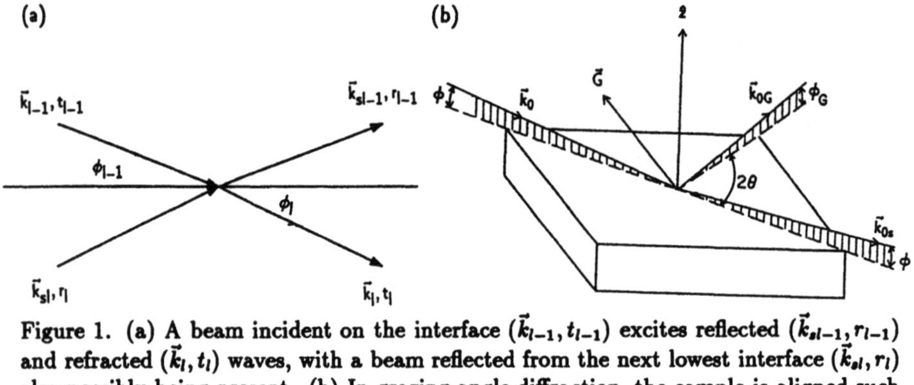

Figure 1. (a) A beam incident on the interface ($\vec{k}_{l-1}, t_{l-1}$) excites reflected ($\vec{k}_{sl-1}, r_{l-1}$) and refracted ($\vec{k}_l, t_l$) waves, with a beam reflected from the next lowest interface ($\vec{k}_{sl}, r_l$) also possibly being present. (b) In grazing-angle diffraction, the sample is aligned such that a diffracted beam ($\vec{k}_{G0}$) is excited simultaneously with the specularly reflected beam.

where

$$R_l = a_l^2 \frac{r_l}{t_l}, \tag{6}$$

$$a_l = e^{ik_{\perp l} d_l / 2}, \tag{7}$$

and

$$F_{l-1,l} = \frac{k_{\perp l-1} - k_{\perp l}}{k_{\perp l-1} + k_{\perp l}}. \tag{8}$$

This recursion relation forms the basis of the modeling schemes commonly employed in reflectivity analysis and the depth-sensitive grazing-angle diffraction formalism described in section 4.

3. Specular Reflectivity Modeling

In a specular reflectivity experiment, one measures the zeroth or vacuum reflectivity $|r_0|^2$. In order to model this data, simply postulate a density profile and apply the recursion relation to it. Starting at the substrate, where $r_L = R_L = 0$, step through the layers until the vacuum, where $t_0 = 1$; r_0 is then uniquely determined (note that in squaring r_0 to compare to the data, all phase information is lost). One then alters this density profile and applies a goodness-of-fit criterion to evaluate the relative merit of different guesses. For example, in our software we minimize

$$\chi^2 = \frac{1}{N - N_f} \sum_{n=1}^{N} \left(\frac{\log_{10} |r_0(Q_n)|^2 - \log_{10} |r_0^{\text{model}}(Q_n)|^2}{u_n / |r_0(Q_n)|^2} \right)^2 \tag{9}$$

using a Levenberg-Marquardt non-linear least squares routine [8], where there are N data points $|r_0(Q_n)|^2$ with experimental uncertainties u_n measured for a discrete set of momentum transfers $Q_n = |\vec{k}_{s0} - \vec{k}_0|_n = (2k_{\perp 0})_n$, with N_f fitted parameters.

The amount of useful information extracted from a given data set depends entirely on the quality of the algorithm used to determine r_0^{model}. For instance, it is rather easy to determine the bilayer spacing of a high-quality multilayer film simply by measuring the positions of its superlattice peaks. However, to obtain information on interfacial mixing requires considerably more effort. A widely used method of modeling interfacial mixing is to simulate a broadened interface by a Debye-Waller factor multiplying the Fresnel coefficients $F_{l-1,l}$ in the recursion formula [4],

$$F_{l-1,l} \Rightarrow F_{l-1,l} e^{-2k_{\perp l-1} k_{\perp l} \sigma_l^2}, \tag{10}$$

where σ_l is the Gaussian width of the region of mixing between layers $l-1$ and l. This approach minimizes the number of slabs L needed to simulate a profile and is rigorously correct in the kinematic limit $\phi_0 \gg \phi_{cl}$. However, there are cases in which the Debye-Waller approach proves inadequate, such as when the interfacial width becomes so large that layer $l-1$ not only mixes with layer l but also with layer $l+1$. A fit routine will quite happily drive the interfacial widths to greater values than this, but the physical meaning of these mixing widths is no longer clear. The correct approach in such a case is to model the system more rigorously by, for example, calculating the interfacial profiles by means of basis functions of the diffusion equation [9] and explicitly generating the interface as a sequence of micro-layers. The micro-layer approach is also necessary when the material does not possess a density profile that can be modeled by macroscopic rectangular slabs with diffuse interfaces, such as when one studies polymers with parabolic segment density profiles [10].

4. Depth-Resolved Grazing-Angle Diffraction

The excitation of a diffracted beam under conditions of total external reflection is known as grazing-angle diffraction. The incident beam strikes the sample at an angle below the critical angle for total external reflection, thereby confining the intensity to an exponentially damped layer at the surface, while the sample is oriented such that the in-plane components of the wavevectors satisfy the Bragg condition [Fig. 1(b)]. In this way, one can use x rays and neutrons to study near-surface structures. Grazing-angle diffraction has most often been employed to determine surface structures using the high intensity available at x-ray synchrotron sources. Some experiments have attempted to study structure beneath the surface, but not in a very quantitative fashion. In this section, I would like to outline how one might apply the well-developed techniques of specular reflectivity modeling to the determination of depth-resolved in-plane structures in grazing-angle diffraction.

Grazing-angle diffraction from imperfect crystals can be described by the distorted wave approximation, which splits the scattering problem into two parts: a kinematical solution of the in-plane Bragg diffraction and a dynamical treatment of the specular reflectivity. In an earlier work [11] we outlined a procedure whereby one might modify the conventional treatment of grazing-angle diffraction from a monolithic slab [12–14] to study the effect of buried layers of different composition and/or structure. The transmittivity and phase factors in that paper were not calculated correctly. In fact, the correct treatment is closely connected with the specular reflectivity formalism discussed above,

$$I(\vec{Q}) = I_0 \left| \sum_l F_l(\vec{Q}_{\parallel}) \delta(\vec{Q}_{\parallel} - \vec{G}_l) t_l(k_{\perp 0}) \frac{e^{iQ_l M_l d_l} - 1}{e^{iQ_l d_l} - 1} t_l(k_{\perp 0 G}) \right|^2 , \tag{11}$$

where the l summation runs over the buried slabs with structure factor F_l, reciprocal lattice vectors G_l, containing M_l layers of width d_l with $\vec{Q}_l = \vec{k}_{Gl} - \vec{k}_l$. The illuminating amplitudes and all of the phase information about the specular process are included in the transmittivity factors, which are given by equations (3)–(8),

$$t_l = \frac{a_l}{a_{l-1}^3} \left(\frac{a_{l-1}^4 - F_{l-1,l} R_{l-1}}{1 - F_{l-1,l}} \right) t_{l-1} . \tag{12}$$

This relationship suggests that any measurement of depth-resolved grazing-angle diffraction must begin with a thorough specular reflectivity characterization in order to obtain the t_l as fixed parameters. One could then study, for example, surface-induced ordering, strain profiles, or magnetization variation with depth, as well as multilayer effects.

5. Prospects

The value of specular reflectivity data depends crucially upon the quality of the models used to simulate it. In many cases, the Debye-Waller formalism provides an adequate representation of the true profile but, in those instances where interdiffusion or roughness is large (for example in many natural oxide layers) or the sample exhibits an unusual density profile, one must explicitly construct the profile from micro-slabs. This procedure is easily within the reach of today's computers. The application of these methods to the calculation of grazing-angle illuminating factors should allow a much more rigorous and quantitative set of measurements to be performed on subsurface depth-dependent ordering. In both cases, one cannot overestimate the importance of accurate modeling in retrieving useful quantitative information from the raw data.

References

[1] C.F. Majkrzak and G.P. Felcher, Mater. Res. Soc. Bull. **15** (11), 65–72 (1990).
[2] J. Penfold and R.K. Thomas, J. Phys. C **2**, 1369 (1990).
[3] T.P. Russell, Mat. Sci. Rep. **5**, 171 (1990).
[4] L. Nevot and P. Croce, Rev. Phys. Appl. **15**, 761 (1980).
[5] F. Abelés, Ann. Phys. Fr. **12**, 596 (1950).
[6] L.G. Parratt, Phys. Rev. **95**, 359 (1954).
[7] H. Rauch and D. Petraschek, in *Neutron Diffraction*, edited by H. Dachs, Topics in Current Physics, Vol. 6 (Springer-Verlag, New York, 1978).
[8] W.H. Press, B.P. Flannery, S.A. Teukolsky, and W.T. Vetterling, *Numerical Recipes* (Cambridge University Press, Cambridge, 1986).
[9] R.W. Erwin, J.J. Rhyne, M.B. Salamon, J. Borchers, Shantanu Sinha, R. Du, J.E. Cunningham, and C.P. Flynn, Phys. Rev. B **35**, 6808 (1987).
[10] S.K. Satija, J.F. Ankner, C.F. Majkrzak, T. Mansfield, D.R. Iyengar, G. Beaucage, T.J. McCarthy, R.S. Stein, and R.J. Composto, to be published.
[11] J.F. Ankner, C.F. Majkrzak, D.A. Neumann, A. Matheny, and C.P. Flynn, Physica B, in press.
[12] G.H. Vineyard, Phys. Rev. B **26**, 4146 (1982).
[13] H. Dosch, B.W. Batterman, and D.C. Wack, Phys. Rev. Lett. **56**, 1144 (1986).
[14] H. Dosch, Phys. Rev. B **35**, 2137 (1987).

Part III

Surface X-Ray Standing Waves

X-Ray Standing Wave Studies of the Liquid/Solid Interface and Ultrathin Organic Films

M.J. Bedzyk

Northwestern University, Department of Materials Science and Engineering, Evanston, IL 60208, USA and
Argonne National Laboratory, Materials Science Division, Argonne, IL 60439, USA

Abstract. X-ray standing waves generated above a mirror surface during total external reflection have been used to locate a heavy atom layer which was embedded hundreds of angstroms above the mirror surface in a Langmuir-Blodgett multilayer. This same method has also been used to map out the ion distribution in the diffuse double layer that forms at the electrolyte / charged surface interface.

1. Introduction

X-rays, which can easily penetrate through matter, are ideal for measuring buried structures, such as interfaces and thin films. However, this property of penetrability makes an x-ray beam inherently less advantageous as a spatially localized probe. To confine the x-ray intensity, one can use the interference fringes (x-ray standing wave) generated from the superposition of two coherently coupled x-ray beams. This gives a spectroscopic probe which is atom specific via the photoeffect and has high spatial resolution perpendicular to the planes of the fringe field. The original method for creating the two interfering beams used dynamical Bragg diffraction [1,2]. More recently, we have used total external reflection [3-6] . In the former case, the standing wave period is equivalent to the diffraction plane spacing. In the latter, the period varies from 10's of angstroms to 100's of angstroms.

2. X-ray Standing Waves Generated by Total External Reflection

Since the index of refraction (n) is less than unity for x-rays in matter, x-rays undergo total external reflection, when the incident angle θ is less than the critical angle θ_c. The index of refraction is expressed as $n = 1 - \delta - i\beta$, where $\delta = N_e r_e \lambda^2/2\pi$ and $\beta = \mu\lambda/4\pi$. Using Snell's law, $\theta_c = (2\delta)^{1/2}$, which is typically between 1 and 10 mrad, depending on the wavelength λ and effective electron density N_e.

During total external reflection, the interference between the incident and specular reflected x-ray plane waves produces an x-ray standing above the mirror surface. The antinodal planes of this periodic E-field intensity pattern are parallel to the surface and have a period of $D = \lambda/2\sin\theta$. Since the total external reflection condition occurs between $\theta = 0$ and $\theta = \theta_c$, the period of the standing wave varies from $D = \infty$ to $D = D_c$. The critical period, $D_c = \lambda/2\theta_c$, is 80 Å for a gold mirror and 200 Å for a silicon mirror. If we ignore anomalous dispersion effects, $D_c = [\pi/(4N_e r_e)]^{1/2}$.

The phase of the standing wave relative to the mirror surface can be determined by using Fresnel theory which describes the magnitude and phase v of the E-field

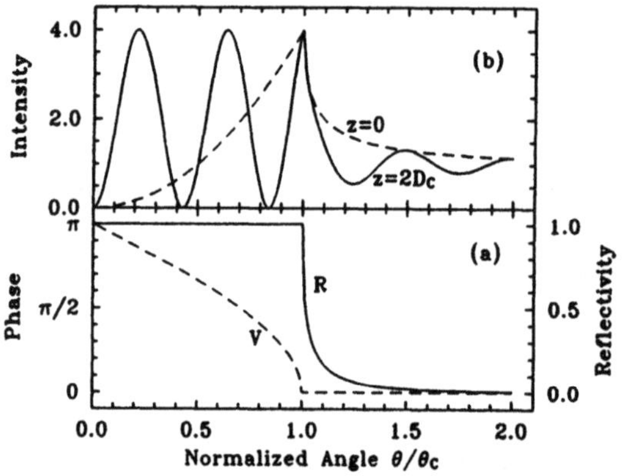

Fig. 1 For $\beta = 0$. (a) The angular dependence of the reflectivity R and relative phase v of the reflected plane wave. (b) The angular dependence of the E-field intensity at $z = 0$ and $z = 2D_c$ for $|E_0| = 1$. (From Ref. 5 .)

amplitude ratio $E_R/E_o = |E_R/E_o|\exp(iv)$. The θ dependence of the reflectivity $R = |E_R/E_o|^2$ and phase v are shown in Fig. 1a. The phase at the mirror surface decreases from π to 0, as the mirror is tilted through the total reflection condition. Thus, at the mirror surface, the reflected plane wave is completely out of phase with the incident plane wave at $\theta = 0$ and completely in phase at $\theta = \theta_c$. Therefore, at $\theta = 0$ a standing wave node is at the mirror surface and the first antinode is at infinity, and as θ is increased the first antinode moves toward the mirror surface, until, it coincides with the mirror surface at $\theta = \theta_c$. As θ increases the second, third, fourth, ... antinodes of the standing wave also move inward, since the period is decreasing as θ increases. The E-field intensity above the mirror surface is expressed as

$$I(\theta,z) = |\mathcal{E}_o + \mathcal{E}_R|^2 = |E_o|^2[1 + R + 2\sqrt{R}\cos(v-2\pi Qz)] ,$$

where $Q = 2\sin\theta/\lambda = 1/D$ is the magnitude of the wave-vector transfer $\mathbf{Q} = \mathbf{k}_R - \mathbf{k}_o$. The total E-field is expressed as the sum of the incident and reflected plane waves $\mathcal{E}_o$ and $\mathcal{E}_R$, respectively. The θ dependence of the intensity $I(\theta, z)$ at $z = 0$ (the mirror surface) and at $z = 2D_c$ is illustrated in Fig. 1b.

Since the photoelectric-effect cross section is proportional to the E-field intensity at the center of an atom, one can determine the position z of an atom layer above the reflecting surface by monitoring the fluorescence yield from that atom layer as the mirror is tilted through the reflection. If the marker layer of atoms is contained in a plane at $z = 2D_c$, the fluorescence yield will have modulations like that shown in Fig. 1b. In general, if an atom layer is at $z = XD_c$, then $X+{}^1/_2$ modulations will occur between $\theta = 0$ and $\theta = \theta_c$.

3. Ultrathin Organic Film Measurement

Such a modulated fluorescence yield was recently observed for the first time [5]. This was for a Zn atom layer embedded in the top arachadate bilayer of a Langmuir-Blodgett multilayer which was deposited on the surface of a gold mirror. As shown

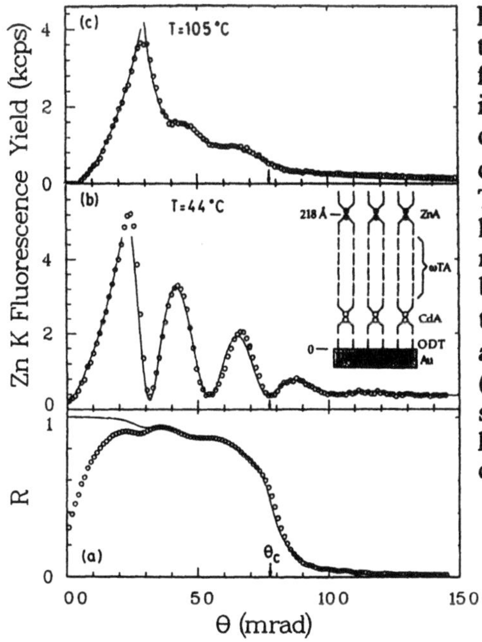

Fig. 2 (a) The angular dependence of the reflectivity at 9.8 keV for the LB film/ Au mirror schematically depicted in the inset. (b) The angular dependence of the Zn Kα fluorescence count rate at T = 44° C, and (c) at T = 105° C. Inset: Circles represent heavy atoms and vertical line segments represent hydrocarbon chains, The LB bilayers are zinc arachadate (ZnA), ω-tricosanoic acid (ωTA), and cadmium arachadate (CdA). Octadecyl thiol (ODT) was adsorbed onto the gold surface to make it uniformly hydrophobic for the subsequent LB depositions. (From Ref. 5 .)

in Fig. 2b, three full oscillations in the Zn fluorescence yield occur before reaching the gold critical angle, indicating that the Zn layer was at $z = 2.5D_C = 200$ Å . This was more precisely determined to be 218 Å by using a χ^2 fit of the data to theoretical yields based on a layered model. From this fit, we also determined that the 2σ thickness of the Zn layer was 24 Å .

Since the standing wave samples the selected atom distribution with a variable period D, we are in effect measuring the Fourier transform of that distribution over a continuous range in Q = 1/D. With a variable period ranging from 10's of angstroms to 100's of angstroms, this new x-ray standing wave probe is ideally suited to measure surface and interface layered structures which have natural length scales in the 10 to 1000 Å regime.

4. Diffuse Double Layer Measurement

We have also directly measured the ionic distribution profile in an electrolyte in contact with a charged phospholipid membrane with long-period x-ray standing waves[6]. In this study, a 27 Å thick membrane was supported from below by a silicon/tungsten layered-synthetic-microstructure (LSM). The x-ray standing waves were generated by total external reflection from the surface of the Si/W LSM. The top of the membrane formed a negatively charged sheet of phosphate ions in contact with a very dilute $ZnCl_2$ aqueous solution. The coverage of the condensed Zn layer (which partially neutralized the phosphate layer) and the Debye length of the electrostatically attracted diffuse Zn ion layer were measured with x-ray standing waves at various pH levels.

The incident x-ray beam, which was tuned to 9.8 keV for optimally exciting Zn K fluorescence, was produced by the 123-pole APS/CHESS undulator [7] at the

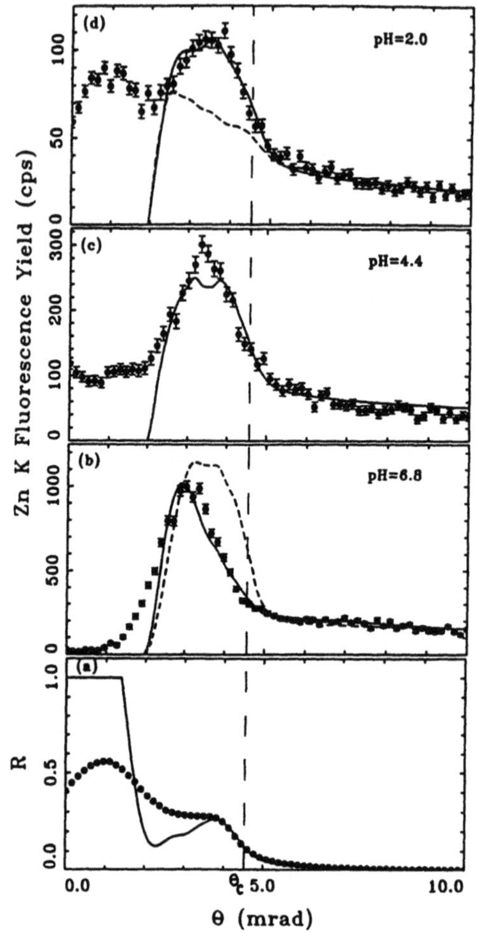

Fig. 3 The experimental (circles) and theoretical (solid lines) angular dependence at $E_\gamma = 9.8$ keV of (a) the specular reflectivity and (b-d) the Zn Ka fluorescence yield for (b) pH 6.8, (c) pH 4.4, and (d) pH 2.0. The solid curves in (b-d) are the χ^2 fit yields for a Zn^{2+} ion distribution which exponentially decays away from the interface. The dashed curve in (b) corresponds to the yield for a discrete Zn^{2+} inner Helmholtz layer at 2 Å above the phosphate headgroup layer. The dashed curve in (d) corresponds to the yield for a flat Zn^{2+} distribution. (From Ref. 6 .)

Cornell Electron Storage Ring (CESR). The data collection consists of recording the reflected intensity with an ion chamber and the fluorescence spectrum with a solid-state detector as a function of angle θ. The simultaneously collected reflectivity and Zn Kα fluorescence data for the initial scan at pH 6.8 are shown in Fig. 3, a and b , respectively.

The solid line curves in Fig. 3 b-d correspond to a χ^2 fit of the data above 2 mrad to a theoretical yield curve which assumes that the Zn^{2+} ions have an exponential decay distribution. The two parameters determined by the fit are the Zn^{2+} concentration in the condensed layer N_c and the Debye length of the exponential decay. We were able to show that changing the pH from 6.8 to 4.4 to 2.0 caused the Debye length to decrease from 58 Å to 8 Å to 3 Å. This represents a weakening of the electrostatic attraction of the charged interface due to protonation of the phosphate head groups of the lipid membrane. Over this same change in pH, N_c went from 0.31 M to 0.31 M to 0.18 M.

5. Conclusion

These results, which are more extensively described in Ref. 6, represent the first detailed structural measurement of the diffuse-double layer. The crucial x-ray properties that made this break-through possible are; (1) the capability of x-rays to penetrate through low-Z materials, such as water, and (2) the capability of using angstrom wavelength x-rays to generate x-ray standing waves with periods ranging from 70 to 1000 Å.

Acknowledgments

My collaborators were Mark Bommarito and Don Bilderback of Cornell University, Martin Caffrey of Ohio State University, and Jay Schildkraut and Tom Penner of the Eastman Kodak Company. This material is based upon work conducted at CHESS, which is supported by the National Science Foundation under award No. DMR 90-21700. This work was also supported in part by a grant from the Eastman Kodak Company.

References

[1] B. W. Batterman, Phys. Rev. **133**, A759 (1964); Phys. Rev. Lett. **22**, 703 (1969).
[2] P. L. Cowan, J. A. Golovchenko, and M. F. Robbins, Phys. Rev. Lett. **44**, 1680 (1980).
[3] M. J. Bedzyk, Nucl. Instr. and Meth. **A266**, 679 (1988).
[4] M. J Bedzyk, D. H. Bilderback, G. M. Bommarito, M. Caffrey, and J. S. Schildkraut, Science **241**, 1788 (1988).
[5] M. J. Bedzyk, G. M. Bommarito, and J. S. Schildkraut, Phys. Rev. Lett. **62**, 1376 (1989).
[6] M. J. Bedzyk, G. M. Bommarito, M. Caffrey, and T. L. Penner, Science **248**, 52 (1990).
[7] D. H. Bilderback et al, Rev. Sci. Instrum. **60**, 1419 (1989).

Glancing-Incidence X-Ray Analysis of Layered Materials

D.K.G. de Boer[1] *and W.W. van den Hoogenhof*[2]

[1]Philips Research Laboratories, P.O. Box 80000, 5600 JA Eindhoven, The Netherlands
[2]Philips Analytical, Lelyweg 1, 7602 EA Almelo, The Netherlands

Abstract. X rays at glancing incidence can set up an X-ray Standing Wave (XSW) in layered materials. In Glancing-Incidence X-ray Analysis (GIXA), the XSW is used as a nanometric yardstick to determine depth distributions of atoms by a combination of reflectivity and x-ray fluorescence. Examples are presented and features resulting from interface roughness are discussed.

If x rays impinge on a flat material under glancing angles of incidence ($\lesssim$ 10 mrad), an X-ray Standing Wave (XSW) can result from interference between incoming and reflected waves. Two types of XSWs are well known in the literature. The first one is that occuring during Bragg reflection from a periodic multilayer [1]; the second is present above a flat surface below the critical angle ψ_c for total reflection [2]. There is a third type, less well known, which is set up in a thin layer (or a few layers) at a reflectivity minimum [3]. The first two mentioned are exploited succesfully [2,4] to determine distances in molecules above solid (or liquid) surfaces.

In this paper it will be shown that XSWs are equally useful for the in-depth analysis of layered materials. The acronym GIXA (Glancing-Incidence X-ray Analysis) is proposed for this technique, a combination of reflectivity and X-Ray Fluorescence (XRF). The experiments shown were obtained using a laboratory equipment, consisting of a conventional diffractometer with an energy-dispersive system for XRF [5]. The theory used is based on a macroscopic optical formalism accounting for absorption and fluorescence enhancement [3]. In all examples, Mo Kα ($\lambda = 0.0711$ nm) is used for excitation. Furthermore, the intensities are normalised at the high-angle value.

Firstly, periodic multilayers will be dealt with. As an example, in fig. 1 a multilayer is considered consisting of 50 periods of 1.6 nm nickel and 4.6 nm carbon on a silicon substrate (which can be used as a soft-x-ray monochromator). The reflectivity exhibits, besides a total-reflection region, peaks due to Bragg diffraction. At diffraction conditions, an XSW exists in (and above) the material. In going from the low-angle side to the high-angle side of the diffraction peak, the antinodes of the standing wave move from the carbon to the nickel layers. Therefore, if XRF signals are measured as a function of the angle of incidence, the Ni-XRF intensity is expected to be minimum at the low-angle side and maximum at the high-angle side, whereas the C XRF follows the reverse behaviour. A small discrepancy between experiment and calculation was

Springer Proceedings in Physics, Vol. 61 **Surface X-Ray and Neutron Scattering**
Eds.: H. Zabel and I.K. Robinson

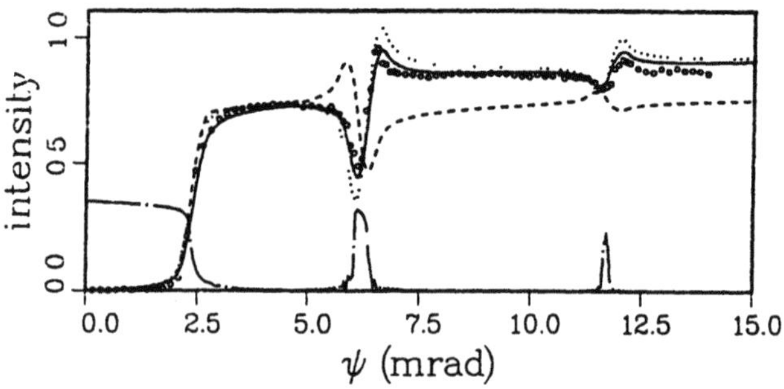

Fig. 1. Intensity *vs* angle of incidence for 50 × (1.6 nm Ni + 4.6 nm C)/Si: (0.35×) reflectivity (dot-dashed); calculated C Kα (dashed); measured Ni Kα (circles), Ni Kα calculated with no Ni in carbon (dotted) and with 10 % Ni dissolved into carbon (solid line). Interface roughness: 0.3 nm for all interfaces.

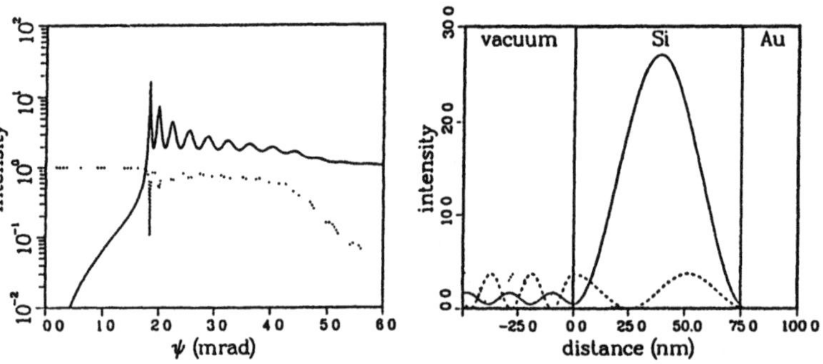

Fig. 2. 75 nm Si on Au. Left: Intensity *vs* angle of incidence ψ: Reflectivity (dotted) and Si Kα (solid line). Right: Relative intensity of exciting x rays *vs* depth at $\psi = 1.6$ (dotted), 1.84 (solid line) and 1.91 mrad (dashed).

found, which was removed by taking into account that *ca.* 10 percent of the nickel had dissolved into the carbon. Thus, the observed angular dependence of glancing-incidence XRF can be used to obtain information on the chemical composition of the various layers in a multilayer.

Whereas the above XSW can be considered as phase modulated (as a function of the incident angle), that above a substrate below ψ_c is both phase and frequency modulated. Now we will discuss the third kind of XSW, which is not only phase and frequency, but also amplitude modulated. As an illustrative example, fig. 2 shows a calculation for a single layer of 75 nm silicon on a gold substrate. Below ψ_c for total reflection at silicon, a standing wave is present above the surface and an evanescent wave inside the material. Above the critical angle, interference phenomena occur, giving rise to Kiessig fringes in the reflectivity. In the silicon layer, besides a beam transmitted through the vacuum-silicon interface, there is a beam reflected at the silicon-gold interface. Moreover, the latter can be reflected again at the silicon-vacuum interface, *etc.*, just like in a wave guide. In a reflectivity minimum, all these waves interfere constructively to yield a standing wave which exactly fits between the two interfaces.

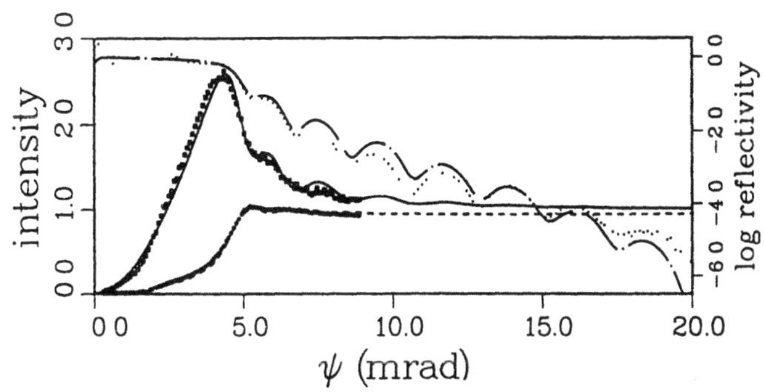

Fig. 3. Intensity *vs* angle of incidence for 0.9 nm Co / 14.3 nm Au / Si: measured (dotted) and calculated (dot-dashed) reflectivity (log scale); measured (squares) and calculated (solid line) Co Kα; measured (circles) and calculated (dashed) Au Lα. Interface roughnesses: 0.5 nm (vacuum/Co), 0.5 nm (Co/Au), 0.7 nm (Au/Si).

This multiple-beam behaviour is characteristic for such a wave-guide-like XSW. The previous two kinds of XSW could be described by two-beam interference, yielding maximum electric fields of twice the incident field and thus maximally four times the incident intensity. The multiple-beam XSW, on the contrary, can be much more intense. This is seen in fig. 2, where the thickness of the silicon layer is chosen in such a way that the effect is large. Note that the depths at which the standing-wave maxima occur, are a function of the angle of incidence. It will be clear that this kind of XSW provides a promising tool for the analysis of thin layers, not only by glancing-incidence XRF, but also by other glancing-incidence x-ray techniqes.

As an example how depth analysis is possible with such XSWs, in fig. 3 a cobalt-gold double layer on silicon is considered. The thickness of the cobalt surface layer is small compared with both evanescent-wave penetration depths and standing-wave periods. Thus, in a reflectivity minimum, where the surface intensity of the wave field is small, the Co XRF intensity has a dip. The Au XRF, on the contrary, has a maximum intensity, because a relatively large amount of radiation can penetrate into the sample. Hence, just like for periodic multilayers, the angular behaviour of glancing-incidence XRF is different for the various layers in layered materials. It is seen in fig. 3 that good agreement between theory and experiment is obtained. For a thinner gold layer (10 nm instead of 14.3 nm) it could be derived that part of the cobalt dissolved into the gold [3].

In the above cases (figs. 1, 3), surface roughness was incorporated in both reflected and transmitted fields [3]. However, GIXA measurements can be complicated by macroscopic roughness, causing reflection in various directions. This is illustrated in fig. 4 for a cobalt-gold (magnetic) multilayer. For the reflectivity, a reasonable fit can be made. In XRF, however, the intensity modulation is much smaller than expected. This can be partly explained by a large intermixing of cobalt and gold. However, a measurement of the rocking curve (*i.e.* a scan of the incident angle with the detector at fixed position) around the first

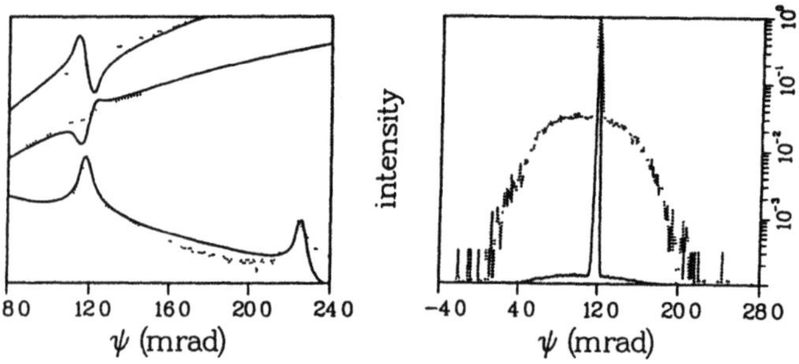

Fig. 4. 100× (1.4 nm Au + 1.8 nm Co) / 14.3 nm Au / Si. Left: Measured (dotted) and calculated (solid lines) intensity *vs* angle of incidence. From top to bottom: Co Kα, Au Lα and log reflectivity (interface roughness: 0.7 nm for all interfaces). Right: Rocking curve measured with detector at the first Bragg peak ($2\psi = 24$ mrad): linear (solid line) and log scale (dotted).

Bragg peak (fig. 4) shows that *ca.* one half of the intensity is not in the specular peak, indicating that large parts of the layered material are not oriented parallel to the substrate surface. This half of the sample does contribute to XRF too, causing a smearing of the angle-dependent structure. Conclusion is that rocking-curve measurements form a nearly indispensable part of GIXA. For the other experimental examples, the rocking curves were neat specular peaks.

Another situation in which roughness plays a role, is that of a granular residue on a flat substrate. This case is encountered in Total-reflection XRF, a method for the chemical analysis of small quantities of material [6] taking advantage of the low background radiation below ψ_c. Now the grains are present in the standing-wave field above the substrate surface and the angular dependence of XRF from a grain will exhibit oscillations. In general, however, there will be a large spread in grain size and hence the oscillations will vanish [7]. Therefore intensities can be summed rather than fields, yielding maximum intensities of twice the incident intensity, indepent of the angle.

In conclusion, from the above examples it will be clear that GIXA with its combination of reflectivity, rocking curves and angle-dependent XRF gives a unique means for non-destructive depth profiling of layered materials.

Acknowledgement. We thank Dr. W. Hoving for providing the cobalt-gold samples.

References

[1] T.W. Barbee Jr. and W.K. Warburton, Mater. Lett. **3**, 17 (1984).
[2] M.J. Bedzyk, G.M. Bommarito, and J.S. Schildkraut, Phys. Rev. Lett. **62**, 1376 (1989).
[3] D.K.G. de Boer, Phys. Rev. B, in press; D.K.G. de Boer and W.W. van den Hoogenhof, Adv. X-Ray Anal. **34**, (1991), in press.

[4] M.J. Bedzyk, D.H. Bilderback, G.M. Bommarito, M. Caffrey and J.S. Schildkraut, Science **241**, 1788 (1988).
[5] D.K.G. de Boer and W.W. van den Hoogenhof, Spectrochim. Acta B, in press.
[6] Proc. 2d TXRF Workshop, Dortmund 1988, Spectrochim. Acta **44B**, No. 5 (1989); Proc. 3d TXRF Workshop, Vienna 1990, Spectrochim. Acta, in press.
[7] D.K.G. de Boer, Spectrochim. Acta B, in press.

Investigation of the Heavy-Atom Distribution in a Langmuir–Blodgett Film by an X-Ray Total External Reflection and Fluorescence Study

S.I. Zheludeva[1], *M.V. Kovalchuk*[1], *N.N. Novikova*[1], *I.V. Bashelhanov*[1], *T. Ishikawa*[2], *and K. Izumi*[2]

[1]A.V. Shubnikov Institute of Crystallography, USSR Academy of Sciences, Leninsky pr. 59, SU-117333 Moscow, USSR
[2]Photon Factory, National Laboratory for High Energy Physics, Oho-machi, Tsukuba-gun, Ibaraki 305, Japan

Abstract. Distributional arrangement of heavy atoms inside a Langmuir-Blodgett (L-B) film was found by means of X-ray fluorescence technique under total external reflection. The thin L-B film (8 monolayers) of stearic acid deposited on silicon substrate was inserted in the solution of lead stearate. The lead fluorescence was measured as a function of the glancing angle of a monochromated(13.5 keV) synchrotron X-ray beam. The analysis indicates that the lead atoms are distributed throughout the L-B film.

L-B films are known to play an important role as optical elements for soft X-ray region, model systems for biomembranes and the main components of molecular electronic devices.

One of advantages of the so-called molecular architecture based on L-B technique is the possibility to control sequence and number of layers and, hence, the characteristics of resulting devices. Another important thing is that using layer by layer deposition you are expected to obtain more sharp interfaces between layers with different properties than in the case of conventional semiconductor technology.

Recently a new technique has been developed for structure characterization of surfaces and interfaces connected with X-ray total external reflection at an optical boundary for glancing angles Θ smaller than critical one.

The interference between incident and specularly reflected beams leads to generation of a standing wave above the mirror surface with period, amplitude and phase strongly dependent on the incident angle. The period of such X-ray standing wave is big enough to be used for determination of organic layers position on the top of metal mirror surface [1], layered synthetic microstructures [2] and thick L-B film [3].

Another important feature of total external reflection is the formation of the so-called evanescent or exponentially damped wave which penetrates a short distance below mirror surface. The investigation of secondary radiation yield excited by an evanescent wave can give information about just the upper surface layers. This idea was realized in the near-total-external- fluorescence technique used for analyses of air-liquid interface [4].

In the present work a complicated multilayer system - a semiconductor mirror with an L-B film on top has been studied under total external reflection conditions.

The sample under investigation - an L-B film (10 mm x 20 mm) containing 8 monolayers of a stearic acid (4 periods of about 50 Å each) deposited on flat hydrophobic Si was inserted twice in the solution of lead stearate $PbSt_2$. In fact, there is a finite possibility for Pb^{++} ions to penetrate inside an acid film and to occupy the head groups of organic molecules. The problem is how many layers of an L-B film have trapped Pb^{++} ions in the process of $PbSt_2$ deposition.

The aim of the work is to find the distributional arrangement of Pb^{++} ions inside the L-B film by registering Pb L_α fluorescence under total external reflection conditions.

The experiment was performed at the Photon Factory on the beam line 15c. Si (111) crystal was used for monochromating the synchrotron X-ray beam. The energy was 13.5 kev, the size of the beam about 40 μm x 4 mm. The intensity of fluorescent X-rays as a function of the glancing angle was measured using Ge solid state detector.

The theoretical calculations are based on the Fresnel-type theory of reflection for a layered structure [5]. A model, in which the unit period of the L-B film is divided into seven layers, have been applied.

For the type of sample studied thin organic film deposited on the solid substrate there are two critical angles Θ_{LB} and Θ_{Si} corresponding to total external reflection conditions for L-B film and Si. Due to lower average electron density of L-B film Θ_{LB} is smaller.

When glancing angle is changed from zero up to Θ_{LB} the fluorescence yield monotonically increases due to the enhancement of evanescent wave penetration depth inside L-B film. For angles greater than Θ_{LB} but still less than Θ_{Si} an incident X-ray beam propagates through L-B film and undergoes total external reflection at the L-B/Si interface. Its interference with specularly reflected beam leads to generation of long-period X-ray standing wave inside L-B film.

E-field intensity depth profiles (fig.1) calculated at fixed angles illustrate the transition from an evanescent wave (curves 1-3) to a standing wave inside L-B film (curves 4-7). The position of nodal planes of X-ray standing wave with respect to the L-B film is strongly determined by the glancing angle, producing the essential modulation of the emission yield from heavy atoms of the L-B film.

The calculated angular dependence of Pb fluorescence yield from Pb^{++} ions positioned in the head groups of organic moleculs at the topmost L-B bilayer is shown in fig.2 (curve 1). A sharp decrease in fluorescence yield is clearly observed at the angle when the first node of the standing wave coincides with the Pb^{++} ions. The greater number of layers are having Pb^{++} ions at the "heads" of organic molecules, the more

Fig.1

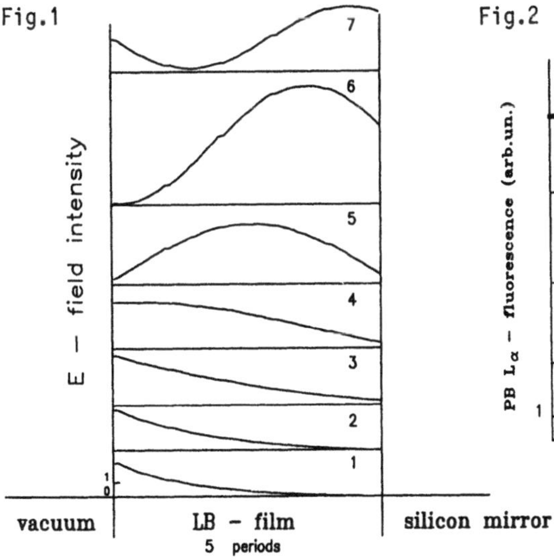

Fig.2

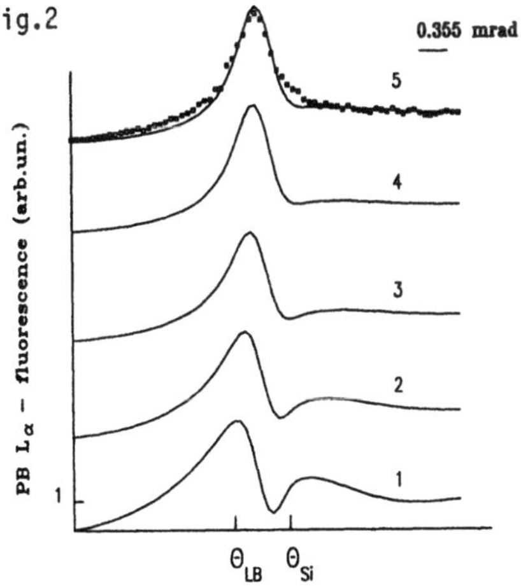

Fig.1. E-field intensity distribution inside an L-B film for increasing glancing angles. Curve (3) corresponds to Θ_{LB}, curve (7) to Θ_{Si}. The angular difference between neighbour curves is about 0.15 mrad.

Fig.2. Calculated angular dependence of Pb L_α fluorescent yield. Curve (1) corresponds to the situation, when Pb^{++} ions are only at the head groups of the upper bilayer; curves 2,3,4,5 - at the "heads" of two, three, four and five bilayers, respectivly. Dotts - experimental results.

transformed will be the fluorescent curve as the thickness of ion distribution becomes comparable with X-ray standing wave period (Fig.2, curves 2-5).

The dotts in Fig.2 represent the experimental results of Pb fluorescence yield and, as it can be seen, are in good agreement with calculated curve 5, which corresponds to the situation, when Pb^{++} ions are at the "heads" of all five organic bilayers composing an L-B film.

As a result it has been stated that Pb^{++} ions penetrate inside a stearic acid film and occupy the "heads" of all the organic layers. In the model used for calculations all the layers possess an equal number of Pb^{++} ions at the head groups.

In conclusion we have demonstrated that total external fluorescence study can be applied to investigation of heavy atom distribution in L-B films. That is of particular importance for characterizing the interface between different layers in complicated multilayer organic structures.

References

[1]M.J.Bedzyk, G.M.Bommarito and J.S.Schildkraut, Phys.Rev.Letters 62,1376(1989)

[2]M.J.Bedzyk, D.N.Bilderback, G.M.Bommarito, M.Caffrey and J.S.Schildkraut, Science **241**,1788(1988)
[3]S.I.Zheludeva, M.V.Kovalchuk, S.Lagomarsino,N.N.Novikova, I.V.Bashelkhanov, V.E.Erochin and L.A.Feigin, Zh.Eksper.Teor.Fiz., Pisma **52**,804(1990)
[4]J.M.Bloch,W.B.Yun, X.Yang, M.Ramanathan, P.A.Montano and C.Capasso, Phys.Rev.Letters **61**,2941(1988)
[5]M.Born and E.Wolf, Principles of Optics(Pergamon Press, New York, 1983).

A Structural Investigation of an Ultra-Thin Langmuir–Blodgett Film by an X-Ray Standing Wave Excited in a LSM Substrate Under the Bragg Diffraction Condition

M.V. Kovalchuk[1], *S.I. Zheludeva*[1], *N.N. Novikova*[1], *I.V. Bashelhanov*[1], *M.J. Bedzyk*[2,*], *and G.M. Bommarito*[3]

[1]A.V. Shubnikov Institute of Crystallography, USSR Academy of Sciences, Leninsky pr. 59, SU-117333 Moscow, USSR
[2]Cornell High Energy Synchrotron Source, Cornell University, Ithaca, NY 14853, USA
[3]Department of Chemistry, Cornell University, Ithaca, NY 14853, USA
*Present address: Department of Materials Science and Engineering, Northwestern University, Evanston, IL 60208, USA

Abstract. An organic multilayer on LSM substrate having the same period was investigated by XSW formed in LSM measuring the fluorescence from heavy atoms of Langmuir-Blodgett(L-B) film. The inclination of organic chains was determined and the interface between the film and LSM was shown to contain additional layer with thickness different for various parts of the sample.

Organic multilayers prepared by the L-B technique are the main elements of molecular electronic devices and therefore demand a thorough investigation of their structural properties. Ultra-thin L-B films comprised of about ten monolayers are of particular interest, since according to the theory of epitaxial development the first few monolayers determine, to a large extend, the structural perfection of thick L-B films. As the electron density, and hence, the scattering efficiency of organic layers is rather low, it is difficult to study ultra-thin film structures by means of conventional X-ray reflection techniques.

A modification of the X-ray diffraction method has been developed, which registers the secondary radiation yield under the diffraction conditions [1]. In a perfect crystal, as a result of dynamic interaction between incident and reflected beams, an X-ray standing wave (XSW) is formed. Its nodes and antinodes move toward the lattice planes as the incident angle is scanned through the Bragg diffraction region. This leads to a drastic change in the photoelectric interaction of X-rays with crystal atoms and, hence, to a peculiar angular dependence of the secondary radiation yield.

Since the standing wave extends above the crystal surface the XSW technique is a powerful tool for structural investigations of near surface layers and overlayer systems.

To investigate the position of an organic bilayer (which is tens of angstrom in thickness) on solid substrate it is necessary to form an XSW with a large period. It can be done by using artificial crystals, namely man-made depth periodic layered synthetic microstructures (LSM) [2] and long period standing waves under total external reflection conditions [3].

If not a bilayer, but a film of a definite thickness is under investigation, the situation becomes much more complicated. The difference between the period of the XSW formed in the substrate and the film d-spacing may destroy the structure-sensitive form of secondary radiation yield. In order to determine the structural parameters of a thin L-B film, XSW with a period very close to organic film d-spacing must be used. In which case, the advantages of using an LSM, which can be prepared with any necessary period, are clear [4,5].

In the present work this idea has been realized for the first time with an ultra-thin organic L-B film on an LSM substrate having the same period and investigated by an XSW formed in the LSM registering the fluorescence from heavy atoms in L-B film.

The sample studied was an L-B film (15mm $\times$ 10mm) of lead behenate (d=59Å) consisting of eleven monolayers. It was deposited on a Rh/C LSM prepared by a laser beam evaporation technique on a Si substrate. (D=59.2Å, D_{Rh}= 10.6Å, the number of periods in the LSM is N=20). The LSM surface was hydrophilic and the existence of an additional layer at the interface of the LSM/L-B film was assumed.

The X-ray reflectivity characterization of the bare LSM has been done using a conventional X-ray tube source at E = 8.04 keV and a Si (111) crystal as monochromator. The beam size was 50 μm $\times$ 3 mm.

The fluorescence yield measurements were done at the Cornell High Energy Synchrotron Source (CHESS) at E = 16.0 keV. The incident beam had a width of 2 mm and a height of 50 μm.

It should be pointed out that at all X-ray experiments and through all angular ranges the dimensions of irradiated sample area were smaller, than that of L-B film.

It is well known that under dynamical Bragg diffraction conditions the E-field intensity distribution in perfect crystals has the form of a standing wave. The quality of XSW and, hence, the sensitivity of the technique depends on the structural perfection of the crystal matrix. Any deviation from the ideal structure will influence the SW pattern. If the crystal used as an XSW generator is sufficiently coherent to give multiple scattering of X-rays, but is far from being perfect, its imperfections must be taken into account to characterize the XSW. The "scale" (XSW) must be carefully studied and calibrated before using.

The preliminary examination of LSM by X-ray reflection in a wide angular range including total external reflection, I and II Bragg orders has been carried out. A good coincidence was obtained between experiment and theoretical calculations based on recursion equations with Frenel coefficients, which took into account such imperfections of LSM as interfacial roughness (3.8Å) and period grading (0.3Å).

The fluorescence data (the upper curves) along with the reflected intensity for the first Bragg order are represented. The deposition of the L-

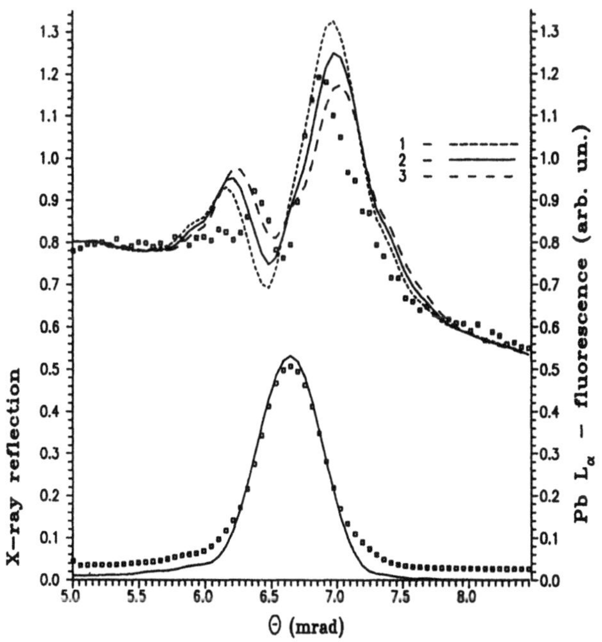

X-ray reflection (bottom curve) and Pb L_α fluorescence intensity. Curves 1,2 and 3 - calculated results for various hydrocarbon chains tilting 27°, 29° and 31°, respectively. Dotts-experimental results.

B film on the LSM did not significantly change X-ray reflection curve in the Bragg diffraction region, because of low electron density and small thickness of organic layer. Under these circumstances the registration of fluorescence signals from heavy atoms within the L-B film is the only way to obtain structural information about the organic layer.

By fitting theory curves based on a Gaussian distribution to the experiment data, information about the mean position and the width of the Gaussian distribution of lead atoms can be obtained. A direct result of the substrate defects influence on the SW pattern is a poor sensitivity of the fluorescence curve to the latter parameter. Thus, the value of the effective width of the Pb layer was taken to be 4Å assuming that in the process of L-B deposition an organic film reproduces LSM surface roughness. Note, that the diameter of Pb^{++} ions ~ 2.4Å.

The position of the heavy atom layers is determined by two varying parameters; the d-spacing of the L-B film and the thickness of some additional layer (water layer, for example) at the interface of the LSM/L-B film. The period of lead atom layers in the direction perpendicular to LSM surface can be less than the length of behenate molecule, because hydrocarbon chains usually are not perpendicular to the substrate, but tilted away from the surface normal.

Curves (1, 2 and 3) represented in figure have been calculated for different L-B film periods 53.8Å, 53.0Å and 52.1Å, which correspond to tilting 27°, 29° and 31°, respectively. They demonstrate that even a 2° variation has a remarkable influence on the form of the curve. The additional layer thickness is zero for that sample area.

Investigation of several parts of the sample showed no differences in the X-ray reflection curves, but fluorescence curves had various forms revealing that the L-B film as well as LSM/L-B interface are not homogeneous. Thus, for one of the sample areas the thickness of additional layer on LSM/L-B interface was estimated to be about 7Å.

Poor coincidence between theory and experiment may be due to the fact that the investigated sample areas are rather large under small glancing angles, their dimensions are about 2mm $\times$ 10mm. Such a region of the L-B film can consist of a large number of microdomains, each with different tilt directions and different LSM/L-B film interface conditions.

In conclusion we have demonstrated that difficulties in structural characterization of ultra-thin organic L-B layers connected with their low scattering efficiency can be overcome with the use of the XSW technique. The main advantage of using an LSM as an XSW generator in this case is the possibility to prepare a SW pattern with the necessary period by varying the LSM parameters.

References

[1]B.W.Batterman,Phys.Rev.133, A759(1964).

[2]M.J.Bedzyk, D.H.Bilderback, G.M.Bommarito, M.Caffrey and J.S.Schildkraut, Science, 247, 1788(1988).

[3]M.J.Bedzyk, G.M.Bommarito and J.S.Schildkraut, Phys.Rev.Lett.62, 1376(1989).

[4]T.W.Barbee and W.K.Warburton, Material Lett.3, 17(1984).

[5]S.I.Zheludeva, M.V.Kovalchuk, N.N.Novikova, I.Yu.Haritonov, A.N.Sosphenov, A.D.Akhsakhaljan, Yu.Ya.Platonov, N.N.Salashchenko Conference Proceedings v.25, II European Conference on Progress of X-ray Synchrotron Radiation Research, Roma 1989 A.Balerna, E.Bernieri and S.Mobilio (Eds) Bologna 1990.

Part IV

Liquid Surfaces

The Structure of Self-Assembled Monolayers

P. Eisenberger[1], *P. Fenter*[1,2], *and K.S. Liang*[2]

[1]Princeton University, Princeton, NJ 08854, USA
[2]Exxon Research and Engineering, Annandale, NJ 08801, USA

Abstract. We have used grazing incidence x-ray diffraction to study the structure of the self-assembled monolayer of $CH_3(CH_2)_{17}SH$ adsorbed on Au(111) and Ag(111). We find that the structure of the monolayer is very different on each of these surfaces. Although on Au(111) the monolayer forms a commensurate $\sqrt{3}x\sqrt{3}R30°$ structure, on Ag(111) the monolayer is both incommensurate and rotated, with a denser lattice spacing.

1. Introduction

In the past few years, there has been a great deal of interest in the properties of organic monolayers adsorbed on solid surfaces, generated by their unique properties as well as their technological importance[1, 2, 3]. A class of monolayers that has been very intensively studied is the self-assembled (SA) monolayer of $CH_3(CH_2)_{17}SH$ (referred to as C_{18}) which form by chemisorption of molecules to a substrate from the *liquid* phase[4, 5, 6, 7]. Because of the structure of the C_{18} *molecule*, the structural arrangement of these molecules in the monolayer is a result of a balance between the competing length and energy scales in the monolayer and the substrate. In the simplest analysis, this creates an attempt to accommodate the packing of the hydrocarbon tails, the sulfur head group spacing and the substrate surface structure. Consequently, a change in one or more of these length scales may produce large changes in the structure of the SA monolayer and substrate surface. For example, these changes may include substrate reconstructions/relaxations, head group structures which are commensurate/incommensurate , and bulk-like/modified hydrocarbon phases. The diversity of structural phenomena observed in this study (even within a given film/substrate combination), suggest the richness of the possible structural rearrangements. This, in turn, clearly indicates the need for coordinated structural and growth studies in order to understand the *intrinsic* properties of these monolayer systems.

Recent work on the SA monolayer of C_{18}/Au(111) has shown that these monolayers have a lattice constant *commensurate* with the substrate[4, 6, 7], although this spacing is ~5% larger than the

Springer Proceedings in Physics, Vol. 61 Surface X-Ray and Neutron Scattering
Eds.: H. Zabel and I.K. Robinson

natural lattice constant of the hydrocarbon tails[8], although this difference can be accommodated by a molecular tilt. By looking at the structure of the same film adsorbed on Au(111) *and* Ag(111) (which have essentially the same lattice constant), we have the opportunity to probe the effect of changing the head group-substrate interaction, leaving other interactions unchanged (to first order).

2. Experimental

Due to the reactivity of the Ag substrate to atmospheric gases, the preparation conditions are important[9] and we will briefly outline them here. These samples were prepared by sputtering and annealing the single crystal substrates under UHV conditions, until the samples were clean and well ordered as measured by Auger electron spectroscopy and low energy electron diffraction. The samples were removed and placed in dilute (~1mM) solutions of 1-octadecanethiol in ethanol in which the films spontaneously form. For the Ag samples, this transfer was done in a dry N_2 environment without exposure to the ambient.

The grazing incidence x-ray diffraction (GIXD) experiments were performed using the Z-axis surface diffractometer on the Exxon X10A beamline at the NSLS using a wavelength of l=1.08Å. The angle of incidence was fixed at 0.4° with respect to the surface plane, and the longitudinal resolution was 0.01$Å^{-1}$. The data shown here were taken in vacuum and the samples were found to be insensitive to beam dose under these conditions.

3. GIXD Results

We have carried out ex-situ GIXD measurements of SA monolayers of C_{18} on both Au(111) and Ag(111) surfaces. Our results demonstrate that there are striking structural differences between films formed on these surfaces (even though they have essentially the same lattice constant). Figure 1a shows a radial scan through the first order diffraction peak of the C_{18} film adsorbed on Au(111) at Q_z=0.21$Å^{-1}$. The data show a clear peak at Q=1.46$Å^{-1}$, with a signal to background of 7/1. The lattice constant of this film (a=4.97Å), as well as its azimuthal orientation (f=30°) demonstrate that this film is commensurate with the underlying Au(111) substrate and forms a √3x√3R30° structure, consistent with previous reports[4, 6, 7]. From the width of this diffraction feature (DQ=0.06$Å^{-1}$), the domain size is found to be (2p/DQ)=105Å.

In constrast, the same film adsorbed on Ag(111) results in a more complex structural arrangement[9]. In figure 1b, a radial scan through the first order diffraction feature is shown. These data

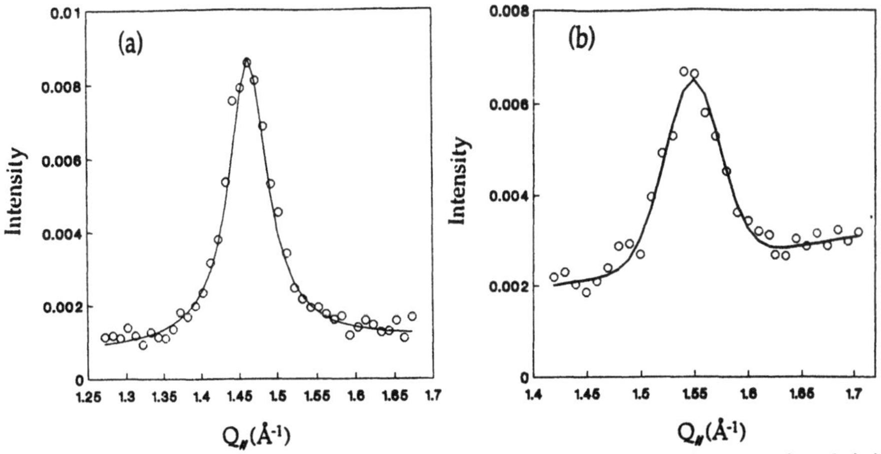

Figure 1. Radial scans through the first order diffraction peak of (a) C_{18}/Au(111), and (b) C_{18}/Ag(111).

demonstrate that the lattice constant of this film (a=4.77Å) is ~5% smaller than the corresponding lattice constant of the C_{18} film on Au(111); that is the monolayer is incommensurate with the surface lattice spacing. In addition, the azimuthal orientation of the film is found to be changed, with the film choosing *two* equivalent domains rotated ±18° from the expected 30° orientation. Lastly, we have observed sharp intense peaks indicative of a Ag surface reconstruction which we believe to be closely related to the monolayer structure.

Interestingly, our studies have also found some variability in the structure of the films on Ag(111)[10]. For instance, a second monolayer was found to be even denser, with a lattice spacing which is ~8% smaller than the commensurate lattice. The monolayer also exhibited the two equivalent domains, but showed an additional splitting which we attribute to an orthorhombic distortion of the hexagonal lattice due to the denser packing.

With the larger 2-d spacing for films adsorbed on Au (as compared to Ag), the molecules are expected to tilt in order to maximize the chain-chain interaction. This 3-d structure of the monolayer can be measured directly by looking at the molecular form factor perpendicular to the surface[7]. A tentative assessment of the the rod profiles has found that the two monolayers exhibit very different 3-d structures[10]. We find that for C_{18}/Au(111) the molecules in the monolayer are tilted by ~30° in the next-nearest neighboring direction, while for C18/Ag(111), the molecules are only slightly tilted by ~8°.

4. Discussion

These data clearly demonstrate that by changing the head group/substrate interaction, a drastic change in the static structural properties can be affected. In the case of C_{18}/Au(111), the monolayer properties (lattice constant, orientation, and tilt angle) are dictated by the substrate lattice spacing in spite of the ~5% mismatch between the $\sqrt{3}$ lattice and the lattice constant of the bulk n-alkanes. In contrast, on Ag(111) the monolayer forms at an incommensurate lattice spacing on a reconstructed surface. The lattice constant and tilt angle of the monolayer are very similar to the bulk n-alkane structure. In both cases the hydrocarbon tails have been able to optimize their interaction, even (in the case of Ag) at the expense of a substrate reconstruction.

5. References

1. J. D. Swalen, D. L. Allara, J. D. Andrade, E. A. Chandross, S. Garoff, J. Israelachvili, T. J. McCarthy, R. Murray, R. F. Pease, J. F. Rabolt, K. J. Wynne, and H. Yu, Langmuir **3**, 932 (1987).
2. G. M. Whitesides, and P. E. Laibinis, Langmuir **6**, 87 (1990).
3. C. E. D. Chidsey, Science **251**, 919 (1991).
4. C. E. D. Chidsey, G. -Y. Liu, P. Rowntree, and G. Scoles, J. Chem. Phys. **91**, 4421 (1989).
5. I. M. Tidswell, T. A. Rabedeau, P. S. Pershan, S. D. Kosowsky, J. P. Folkers, and G. M. Whitesides, unpublished.
6. L. Strong, and G. M. Whitesides, Langmuir **4**, 546 (1988).
7. M. G. Samant, C. A. Brown, and J. G. Gordon II, Langmuir **7**, 437 (1991).
8. J. Doucet, I. Denicolo, and A. Craievich, J. Chem. Phys. **75**, 1523 (1981).
9. P. Fenter, P. Eisenberger, J. Li, N. Camillone III, S. Bernasek, G. Scoles, T. A. Ramanarayanan, and K. S. Liang, submitted to Langmuir.
10. P. Fenter et al, unpublished.

Behenic Acid as a Structural Model for Fatty Acid Monolayers at the Air/Water Interface: An X-Ray Diffraction Study

R.M. Kenn[1], *C. Böhm*[1], *H. Möhwald*[1], *K. Kjaer*[2], and *J. Als-Nielsen*[2]

[1]Universität Mainz, Department of Physical Chemistry, W-6500 Mainz, Fed. Rep. of Germany
[2]Risø National Laboratory, DK-4000 Roskilde, Denmark

We present an X-ray diffraction study of behenic (docosanoic) acid films at the air/water interface. Analysis of the rod profiles parallel and perpendicular to the surface provides detailed information on the unit cells of five different phases.

2 Introduction

Monolayers of fatty acids ($CH_3(CH_2)_{n-2}COOH$) at the air/water interface have been studied intensively for many years by isotherm experiments, and in recent time also by X-ray diffraction. From the Π vs. A measurments (Π, A, T, n) phase diagrams have been deduced [1, 2] which, together with X-ray results [3, 4, 5, 6], allow the conclusion that the (Π, A, T) phase diagrams are similiar for the homologous series, with longer chain length n equivalent to lower temperature.

3 Results

In Fig. 1 the (Π, T) phase diagram for behenic acid ($n = 22$) monolayers is shown. The x-ray data lead to the following structural interpretation: we find

- 4 ordered orthorhombic phases:
  - two of them with uniform tilt in symmetry directions (L_2: towards nearest, L_2': towards next nearest neighbour) and short range positional coherence length;
  - the other two with zero tilt, but different peak widths, indicating different positional coherence lengths; (CS: long range positional order, crystalline packed; S: only short range positional order)
- 1 ordered hexagonal phase with vertical chain alignment and short range positional order (LS).

In Fig. 2 the linewidth of the different reflections observed is plotted against the lateral pressure. Clearly an anisotropy in the linewidths is present in the L_2-phase, in the L_2'-phase it is less pronounced but still present in the S-phase.

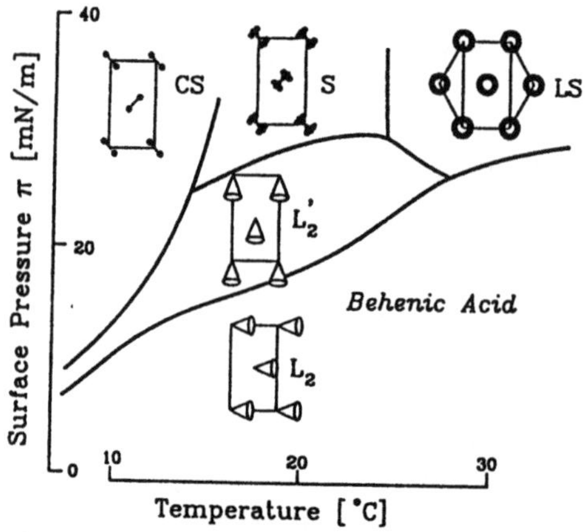

Figure 1: Phase diagram of behenic acid on water, as derived by isotherms [1]. The schematics show the structures of the phases as deduced from x-ray diffraction [3].

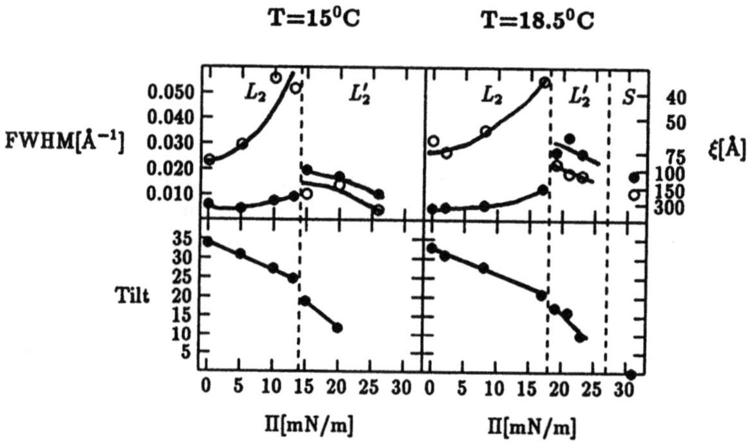

Figure 2: Top: Line widths vs. lateral pressure at two different temperatures. The open circles correspond to the {11}–reflection, the closed circles to the {20}–reflection. Bottom: Tilt angle vs. lateral pressure. The solid lines are a guide to the eye.

The anisotropy of the widths of the reflections is coupled to the molecular tilt direction and magnitude. In the L_2'–phase, a tilt towards the nearest neighbour is observed, because the {11} reflections occur at $Q_z > Q_c$, Q_c being the wavevector transfer at the critical angle, whereas the {20}–reflection peaks at $Q_z = Q_c$[3]. Here, the width of the {11}–reflection is clearly larger than that of the {20}–reflection. In the L_2'–phase, both reflections occur at $Q_z > Q_c$, with $Q_z^{\{20\}} > Q_z^{\{11\}}$. Thus it seems likely, that the anisotropy of the widths depends not only on the tilt magnitude, but also on the tilt azimuth, or more precisely, on the angle between the tilt direction and the reciprocal lattice vector. This angle is in the L_2–phase smaller for the {11}–reflection than for the {20}–reflection, whereas in the L_2'–phase the situation is interchanged.

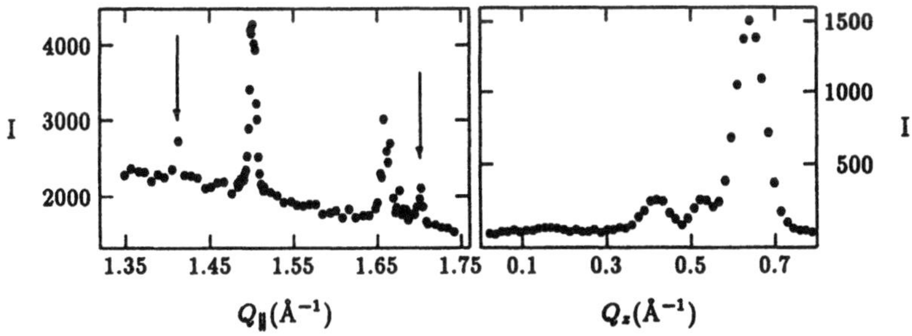

Figure 3: X-ray diffraction pattern of behenic acid on water after the area relaxation has occured. Left: Intensity vs. in-plane wavevector transfer. The peaks marked with an arrow are due to the collapsed moiety of the film. Right: rod scan taken at the position of the left marked peak. Note: intensities on different scales.

The coherence length *decreases* in the L_2–phase with decreasing tilt, whereas it *increases* in the L_2'–phase with decreasing tilt. The mechanism for this is not yet completely understood [3], but together with the observation of defect annealing at the L_2–L_2'–phase boundary [7] a hint for this can be given: upon compression in the L_2–phase, a part of the molecules is locked at a lower tilt angle, thus inducing defects. Crossing the phase boundary, the tilt azimuth changes [3], thus inducing a softening of the lattice. This is now a possibility for the locked molecules to acquire the right tilt angle, which reduces the number of defects. Of course, this can not explain the decrease of the linewidth upon compression in the L_2'–phase.

It has been shown (cf. [8] and literature citted herein) that fatty acid monolayers are in the best case metastable and show a transition into a multilayer phase. This instability is recognized by a loss of film area when held at constant pressure. The onset of this instability changes the diffraction pattern drastically, as shown in Fig. 3. Two supplementary peaks, marked by an arrow, occur. The shape of their bragg rods is not the expexted for a monolayer. From the widths of these rod profiles, the thickness of the scatterer can be estimated to be ≈ 75Å. Comparing this with 30 Å length for behenic acid allows the conclusion, that the first step of the multilayer formation occurs via the formation of a trilayer.

The facts mentioned above support the notion that fatty acid monolayers exhibit mesophases which can be compared to usual categories of thermotropic liquid crystals. The phases L_2, L_2', LS and S show exactly the same unit cell symmetries as the smectic phases I, F, E and B respectively. The absence of long range positional correlations has been proved by X-ray diffraction [3], whereas the presence of long range orientational order has been confirmed by fluorescence microscopy for the L_2–phase [9] and indirectly by polarisation microscopy for the L_2', LS and S–phase [10].

Acknowledgment. Thanks are due to the staff at HASYLAB for ample beamtime and support. This work was funded by the Bundesministerium für Forschung und Technologie (BMFT), the Deutsche Forschungsgemeinschaft, the Materialwissenschaftliche Forschungszentrum of the Universität Mainz and the Danish Natural Science Foundation.

References

[1] A. M. Bibo and I. R. Peterson, *Adv. Mat.*, **2**, 309 (1990).
[2] S. Ställberg-Stenhagen and E. Stenhagen, *Nature*, **156**, 239 (1945).
[3] R. M. Kenn, C. Böhm, A. M. Bibo, I. R. Petersen, H. Möhwald, K. Kjaer and J. Als-Nielsen, *J. Phys. Chem.*, **95**, 2092 (1991).
[4] K. Kjaer, J. Als-Nielsen, C. A. Helm, P. Tippmann-Krayer and H. Möhwald, *J. Phys. Chem.*, **93**, 3200 (1989).
[5] B. Lin, M. C. Shih, T. M. Bohanon, G. E. Ice and P. Dutta, *Phys. Rev. Lett.*, **65**, 191 (1990).
[6] M. L. Schlossmann, D. K. Schwartz, P. S. Pershan, E. H. Kawamoto, G. J. Kellogg and S. Lee, *Phys. Rev. Lett.*, **66**, 1599 (1991).
[7] A. M. Bibo and I. R. Peterson, *Thin Solid Films*, **178**, 81 (1989).
[8] E. Pezron, P. M. Claesson, J. M. Berg and D. Vollhardt, *J. Coll. Interf. Sci.*, **138**, 245 (1989).
[9] V. T. Moy, D. Keller, H. E. Gaub and H. M. McConnel, *J. Phys. Chem.*, **90**, 3198 (1986).
[10] G. Veale, I.R Girling and I. R. Peterson, *Thin Solid Films*, **127**, 293 (1985).

X-Ray Scattering Studies of Organic Monolayers on Electrolytic Solutions: Arachidic Acid on $CdCl_2$

K. Kjaer[1], J. Als-Nielsen[1], R.M. Kenn[2], C. Böhm[2], P. Tippmann-Krayer[2], C.A. Helm[2], H. Möhwald[2], F. Leveiller[3], D. Jacquemain[3], M. Lahav[3], L. Leiserowitz[3], and M. Deutsch[4]

[1]Physics Department, Risø National Laboratory, DK-4000 Roskilde, Denmark
[2]Physical Chemistry Department, Mainz University, W-6500 Mainz, Fed. Rep. of Germany
[3]Structural Chemistry Department, Weizmann Institute, Rehovot, 76100 Israel
[4]Physics Department, Bar-Ilan University, Ramat-Gan, 52100 Israel

Abstract. The interaction between a charged monolayer of fatty acid molecules on the surface of a $CdCl_2$ aqueous solution and the ions below has been investigated by means of X-ray Reflection and Grazing-Incidence Diffraction. A stoichiometric, localised layer of Cd^{++} ions forms an epitaxial 2×3 superstructure below the 2D-crystalline fatty acid monolayer.

1. Introduction

Amphiphilic organic molecules, when deposited on the surface of water, may form monolayers: so-called *Langmuir films*. These quasi-2D systems may be manipulated by variation of mean area per molecule A, surface pressure π, temperature etc. [1]. For a charged monolayer, the counter ions attracted from the subphase may be bound and localised or they may form a diffuse layer.

2. Arachidic acid $CH_3(CH_2)_{18}COOH$ on $CdCl_2$ solutions

Room temperature X-ray specular reflection data for this system [2] are shown in figure 1a which compares data for $pH \simeq 5.6$, $A = 19.8$ Å^2 and $\pi \simeq 25$ mN/m (lower curve) with data for films on dilute $CdCl_2$ solutions at $A = 20.1$ Å^2 and $\pi \simeq 0$ mN/m and $pH \simeq 7$ [3]. The data show that the charged monolayer interacts strongly with the ionic subphase, and that the effect is nearly saturated even at 10^{-5} M $CdCl_2$. The data were interpreted in terms of slab models [5] of the vertical electron density profiles $\rho(z)$ as shown in figure 1b. For the film on pure water, $\rho(z)$ (full curve) has 32 electrons per molecule in a 3.1 Å thick head group slab, consistent with one COOH plus one H_2O. The profile for 10^{-3} M $CdCl_2$ has 53 electrons per molecule in a 3.0 Å thick head group slab, consistent with the addition of 0.46 Cd^{++}, in good agreement with the expected stoichiometry of $\frac{1}{2}Cd^{++}$ per molecule.

Figure 2b shows Grazing-Incidence Diffraction data [6] for Arachidic acid monolayers at $\pi \simeq 0$ ($A = 28$ Å^2 per molecule) on a subphase of 10^{-3} M $CdCl_2$ and $pH \simeq 8.8$ (adjusted with ammonia [7]). The diffraction pattern (figure 2b) appeared only after cooling from room temperature to $T \simeq 9$ °C. The posi-

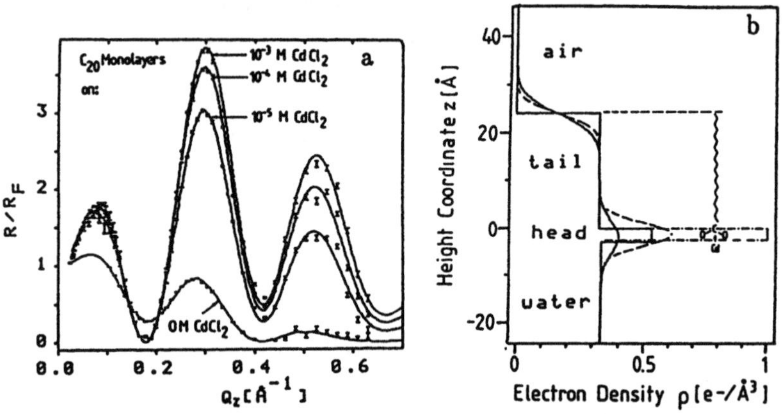

Figure 1: a: Normalised specular reflectivity data (R/R_F *vs.* q_z) for Arachidic acid on pure water (lower curve) and on $CdCl_2$ solutions. b: Fitted vertical density profiles $\rho(z)$. The smooth curves include the effect of roughness [5]. Full curve: monolayer on pure water; dashed curve: monolayer on 10^{-3} M $CdCl_2$.

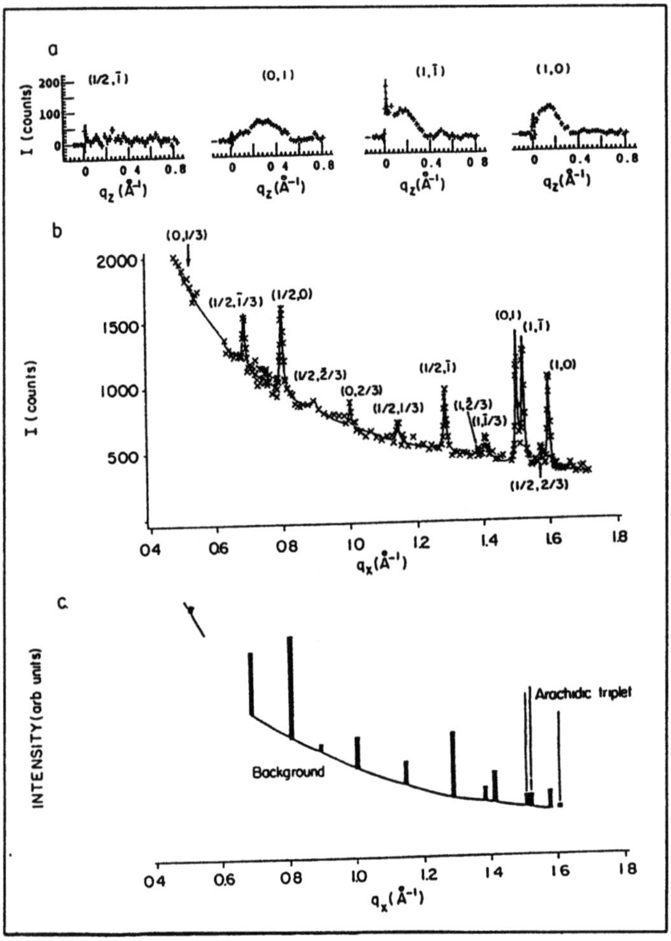

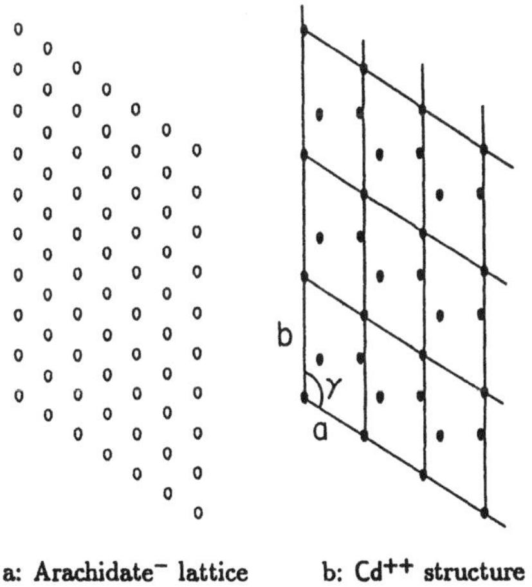

Figure 3: a: Arachidate$^-$ lattice, and
b: Cd^{++} positions in the 2×3 epitaxial structure.

tional coherence is of long range ($\simeq$ 1000 Å). To interpret these data, consider first the Bragg Rods (figure 2a) of the triplet of peaks near $q_x \simeq 1.55$ Å^{-1}. The sharp peak near $q_z \simeq 0.01$ Å^{-1} is a manifestation of the factor $|T(\alpha_f)|^2$ [8] which results from the Bragg Rod interfering constructively with its mirror image total-reflected at the interface. The width of the broad features is FWHM(q_z) $\simeq 2\pi/(25$ Å), showing that the triplet is due to the $\simeq$25 Å long amphiphile molecules, ordered in an oblique 2D-crystal structure (figure 3a). The Bragg Rods indicate that the molecules are uniformly tilted by $11 \pm 1°$ in a next-nearest-neighbour direction (to within $\pm 5°$ [6,9]). The remaining peaks have flat Bragg Rods to beyond 0.8 Å^{-1} (figure 2a), showing that they stem from a structure <3-4 Å thick, consistent with the source of the scattering being the attracted Cd^{++} layer [6]. The peaks index as $(h/2, k/3)$ relative to the triplet, indicating a commensurate 2×3 super-structure. The positions of three Cd^{++} in the 2×3 cell were refined [6,10] to a unique model (figure 3b), giving calculated intensities (figure 2c) which agree well with the data (figure 2b).

3. Conclusion

To conclude, specular reflectivity data show extra density in a thin layer at or just below the head groups of the ionised monolayer floating on dilute $CdCl_2$ solutions. The extra density is consistent with a layer of Cd^{++} in the expected stoichiometry. Analysis of the Bragg Rods observed in Grazing-Incidence Diffraction allows the conclusion that the triplet of integer-order peaks stems from a 2D-crystalline structure involving mainly the Arachidate molecules while the source of the remaining peaks is a thin 2D-crystal. We

Figure 2: a: Bragg rods (I vs. q_z), and
b: Horizontal diffraction data (I vs. q_x), for Arachidic acid on $CdCl_2$ solutions.
c: Calculated intensities for the fitted model (figure 3).

have concluded that the main constituent of this thin crystalline layer is the Cd^{++} ions attracted from the subphase by the charged monolayer. The Cd^{++} crystal has an epitaxial relationship to the Arachidate structure. The Bragg Rods give the orientation of the surfactant molecules to within $\pm 1°$ for the tilt angle from vertical and $\pm 5°$ for the azimuthal angle.

Work is in progress to deduce the registry of the Cd^{++} lattices by further analysis of the integer-order Bragg Rods, and on determining the rôle (if any) of NH_3 [7] and H_2O in the 2×3 structure by comparison with bases other than NH_3 and ions other than Cd^{++}. Anomalous scattering and Near-Total External Fluorescence studies may prove to be of use as well.

We are indebted to HASYLAB at DESY, Hamburg for synchrotron beam time and support. This work was funded by the Bundesministerium für Forschung und Technologie (BMFT), the Deutsche Forschungsgemeinschaft (SFB262), the Materialwissenschaftliche Forschungszentrum (MWFZ) of the Univ. Mainz, the Fonds der Chemischen Industrie, the U.S./Israel Binational Science Foundation, the Petroleum Fund of the American Chemical Society, the German-Israeli Foundation, the Fund for Basic Research of the Israeli Academy of Sciences and the Danish Natural Science Foundation.

References

[1] G. L. Gaines, "Insoluble Monolayers at Liquid-Gas Interfaces"; Interscience: New York (1966).

[2] K. Kjaer, J. Als-Nielsen, C. A. Helm, P. Tippmann-Krayer and H. Möhwald, *J. Phys. Chem.* **93**, pp. 3200–6 (1989).

[3] In ref. [2], *pH* was adjusted with NaOH and could not be determined very accurately. We estimate now that *pH* was about 7, high enough to completely dissociate COOH to COO^-. At higher *pH*, precipitation occurs [4].

[4] *Gmelins Handbuch der Anorganischen Chemie*, **33**:Cadmium, p. 436. Eighth edition (1959).

[5] J. Als-Nielsen and K. Kjaer, In: "Phase Transitions in Soft Condensed Matter"; T. Riste and D. Sherrington, Eds; *Proc. NATO ASI* Series B 211, Plenum: New York, pp. 113–38 (1989).

[6] F. Leveiller, D. Jacquemain, M. Lahav, L. Leiserowitz, M. Deutsch, K. Kjaer and J. Als-Nielsen, *Science* **252**, pp. 1532–6 (1991).

[7] *pH* was adjusted with NH_3 and measured to within 0.1. The use of NH_3 allows $pH \leq 8.8$ due to the formation of the complexe $Cd(NH_3)_4^{++}$ (cf. ref. [4], p.443). At higher *pH* we observed the precipitation of a fine white powder.

[8] Cf. G. Vineyard, *Phys. Rev. B* **26**, pp. 4146–59 (1982) and papers in this volume.

[9] F. Leveiller *et al.*, to be published (1991).

[10] Five parameters were adjusted to fit ten observed intensities.

The Phases of Phosphatidyl Ethanolamine Monolayers

C.A. Helm[1], *P. Tippmann-Krayer*[1], *R.M. Kenn*[1], *H. Möhwald*[1], *J. Als-Nielsen*[2], and *K. Kjaer*[2]

[1]Department of Physical Chemistry, Johannes Gutenberg Universität Mainz, W-6500 Mainz, Fed. Rep. of Germany
[2]Physics Department, Risø National Laboratory, DK-4000 Roskilde, Denmark

Abstract For the first time phospholipid monolayers at the air/water interface have been studied by X-ray diffraction and reflection along the whole isotherm from the isotropic fluid to the ordered phases [1]. The model used to analyze the data - and the accuracy of the parameters deduced - were tested by comparing the results obtained with two lipids having the same head group but different chain lengths.

1. Introduction

A lipid molecule has two very distinctive regions: a long hydrocarbon tail which repels water and a polar or charged head which strongly attracts water. A small amount of lipid will spread over the surface of water to form a *monolayer*; in this thin film the tail regions face the air and the head groups are in close contact with water. Two such films can combine tail to tail to make a phospholipid sandwich or *lipid bilayer*, which is the structural basis of all cell membranes.

2. Results and Analysis

Two different scattering geometries were employed [2]. Specular reflection measures the x-ray reflectivity perpendicular to the interface, which would be described by the Fresnel law for an infinitely thin interface. For a real monolayer the deviation of the measured reflectivity R from the Fresnel prediction R_F gives information about the electron density profile perpendicular to the interface. A lipid monolayer can be modeled as consisting of two regions of constant electron density: one box for the tail section (electron density ρ_{tail}, vertical extension l_{tail}) and one for the head group region (ρ_{head}, l_{head}) [3,4]. Diffraction at grazing incidence probes the structure within the interface, and for lateral positional order, Bragg peaks result [5].
The inset of Fig. 1 gives the isotherm (surface pressure π vs. molecular area A) of a typical lipid monolayer [6], in this case DMPE (L-α-Dimyristoyl-Phosphatidylethanolamine). On increase of the surface pressure an isotropic *fluid phase* appears. The reflectivity curve (Fig.1) has a relatively shallow minimum which moves to lower values of Q_z on compression. Detailed analysis shows that compression leads predominantly to a change of thickness of the hydrophobic moiety, much less of its density, with the head group extension remaining constant (Fig. 2). No lateral order is observed.
As π exceeds a certain lateral pressure the isotherm becomes nearly horizontal, corresponding to the onset of the so-called *coexistence region*. In this regime, the fluid and an ordered phase coexist and the area fraction of the latter is increased on compression. Incoherent superposition of the reflectivity curves according to the area

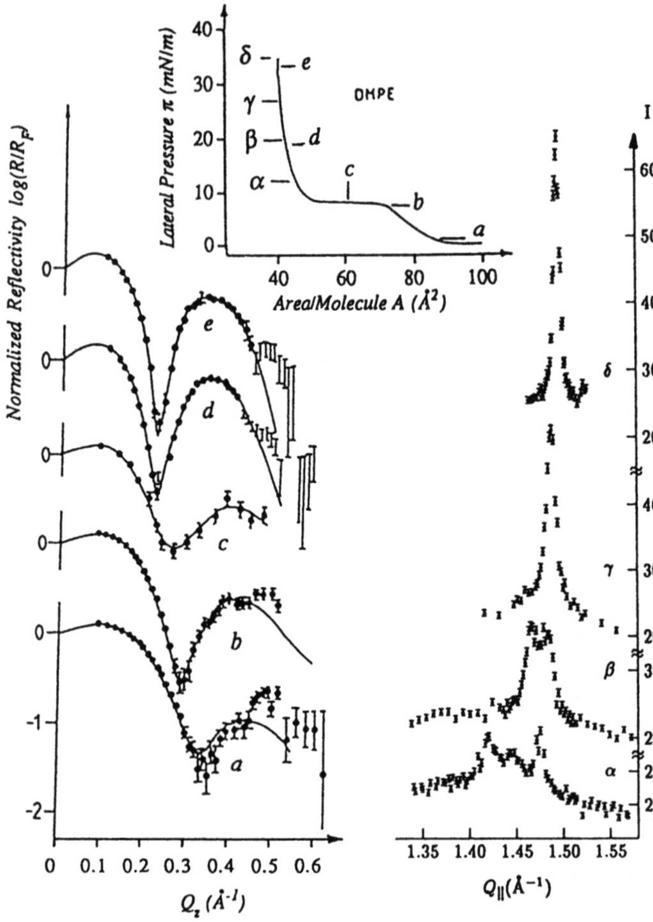

Fig. 1 Left: Reflectivity curves *vs.* vertical scattering vector Q_z for DMPE in the isotropic fluid phase (a and b), in the coexisting region (c) and in the condensed phase (e and d). For clarity, each curve is displaced one decade. Right: In-plane diffraction data for DMPE in the ordered phases. Inset: Isotherm.

fraction of the coexisting phases can explain the broadening of the minimum (Fig.1, left). This phase transition involves a considerable increase (≈10%) of the electron density in the hydrophobic region, a dehydration of the head group and a positional ordering of the aliphatic tails (Fig.1, right), albeit with low coherence lengths (≈10 spacings).

On further compression of the film, the *ordered phase* undergoes a continuous transition. This is characterized by an increase in positional ordering (cf. Fig.1, right), a discontinuous decrease in lateral compressibility, a decrease in chain tilt angle with respect to the surface normal towards zero and also a head group dehydration.

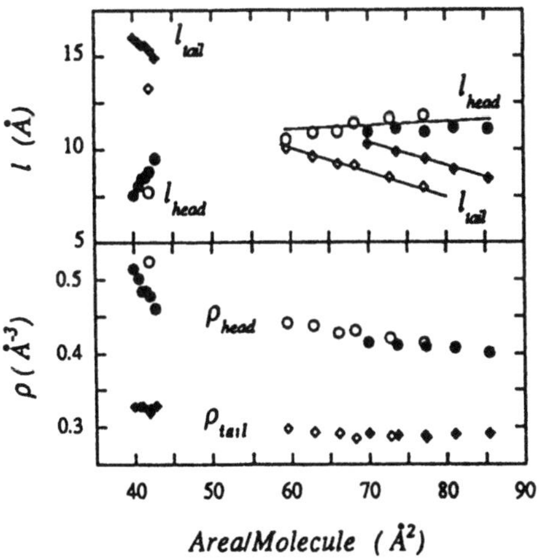

Fig. 2
Structural parameters of monolayers of DMPE (filled symbols) and DLPE (open symbols) - the only difference is the number of CH_2 groups in the tails - as deduced from reflectivity curves *vs.* mean molecular area *A*. The left part shows the ordered phases and the right part the fluid phase.

3. References

[1] C.A.Helm, P. Tippmann-Krayer, H.Möhwald, K.Kjaer and J.Als-Nielsen *Biophys. J.*, in print.

[2] J.Als-Nielsen and K.Kjaer in "Phase Transitions in Soft Condensed Matter" (Eds: T.Riste and D.Sherrington). Plenum Press (1989).

[3] C.A.Helm, H.Möhwald, K.Kjaer and J.Als-Nielsen *Europhys. Lett*, **4**, 697 (1987).

[4] J.Daillant, L.Bosio, J.J.Benattar, J.Meunier, *Europhys. Lett*, **8**, 453 (1989).

[5] C.A.Helm, H.Möhwald, K.Kjaer and J. Als-Nielsen *Biophys. J.* **52**, 381 (1987).

[6] O.Albrecht, H.Gruler and E.Sackmann *J. Phys. (Paris)*, **39**, 301 (1978).

X-Ray Diffraction Studies of Fatty Acid Monolayers on the Surface of Water

M.C. Shih[1], *T.M. Bohanon*[1], *J.M. Mikrut*[1], *P. Zschack*[2], and *P. Dutta*[1]

[1]Department of Physics and Astronomy, Northwestern University, Evanston, IL 60208, USA

[2]Metals and Ceramics Division, Oak Ridge National Laboratory, Oak Ridge, TN 37830, USA

Abstract. We review the current status of X-ray diffraction studies of Langmuir films. The structures of most known fatty acid monolayer phases have now been determined (and found to be different from what had historically been assumed); and various previously unexplained observations have now been given a microscopic basis.

For much of the history of Langmuir monolayer studies, people have tended to assume that if two isotherms have the same features, the phases separated by these features are the same in the two cases. Also, many monolayer studies are still performed only at poorly-controlled room temperature. However, Stenhagen [1] and Lundquist [2] found decades ago that, if isotherms are recorded as a function of temperature, they indicate quite complex phase diagrams. Their work has been largely ignored; it is only recently that Bibo et al. [3,4] have reminded the community of the existence of the earlier work.

Of course, isotherm studies do not tell us what the phases are; this requires synchrotron radiation. Lin et al. [5] determined the structures of four heneicosanoic acid phases in 1990; two other structures were reported soon afterwards in heneicosanoic acid [6] and also in docosanoic acid [7]. Since the only known effect of chain length is to shift the phase diagram [3], a generalized fatty acid phase diagram can be drawn (Fig. 1; cf. Ref. 4). The structures of six of the phases in the figure were determined earlier, as mentioned above, while the remaining two phases (Rotator-I, C') are indicated by our recent, unpublished diffraction data [8].

We will first describe the six previously-known phases. These phases have historically been labelled L_2 ('liquid no. 2'), LS ('superliquid'), etc., but such assumptions are now known to be incorrect or incomplete. Phases A and A' have vertical chains, and the lattice is distorted hexagonal; they differ only slightly in their lattice spacings [5]. At higher temperatures, the lattice becomes hexagonal [6,7]; we call this phase Rotator-II after the similar phase seen in lamellar paraffins. B, C and D all have distorted-hexagonal in-plane lattices and tilted chains [5-7]. In B and D, the tilt is towards a near neighbor, but the lattice spacings are different. C has a chain tilt direction halfway between two near neighbors.

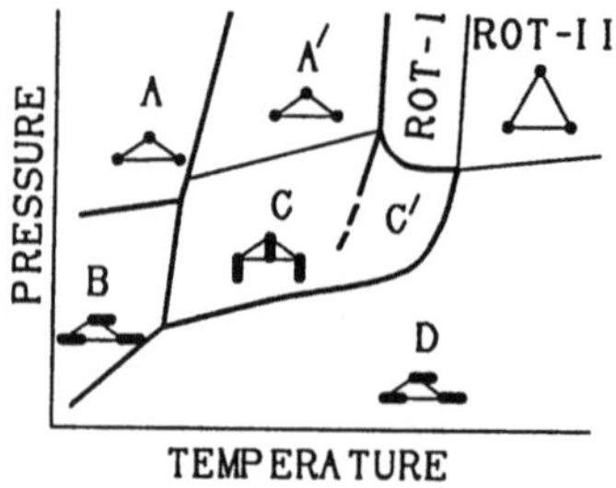

Fig. 1: *Qualitative* phase diagram (bold line = first order, thin line = continuous, dashed line = uncertain). In schematics, molecules are represented by rods projected into the horizontal plane, so that elongation indicates tilt direction; distortion of triangles is exaggerated. The A and A' structures differ only in their lattice spacings; the same is true for C and C', and for B and D. Rotator-I has vertical chains and distortion intermediate between A' and Rotator-II (see text). The traditional names for these phases (see e.g. Ref. 1-4) are A = CS (condensed solid), A' = S (solid); Rotator-II = LS (superliquid); D, C, B = L_2, L_2', L_2'' (liquids). In fact all these phases are at least partially ordered.

These phase identifications, while well-established, do not explain a well-known phenomenon. The highest-pressure discontinuity in the isotherm, which indicates the onset of tilt, goes from a kink to a flat section and back to a kink within a small temperature range. (This behavior is well known mainly because the monolayer mechanical properties are anomalous in this region; see below.) We have found, from X-ray diffraction, that the A' and Rotator-II phases are separated by a 'Rotator-I' phase [8]. This phase, like the bulk paraffin phase after which it is named but unlike the neighboring A' and Rotator-II phases, shows a strongly temperature-dependent unit cell area and aspect ratio (b/a). The transition marking the onset of tilt is a flat section only where this new phase occurs. Below the flat section, we see another new phase [8]; it is similar to C but has slightly different lattice spacings, so we have labelled it C'. Although most isotherms show no sign of a C/C' transition, a new (and very small) isotherm feature in the right place was reported very recently by Bibo et al. [4]. The convergence of a number of phase boundaries in this small region allows us to explain the otherwise peculiar behavior of the isotherm feature.

Incidentally, the 'reentrant appearance of phases', time dependence, and 'bizarre coexistence' reported by Schlossmann et al. [9] for tetracosanoic acid along a room temperature isotherm can be understood with reference to Fig. 1. It is possible to go along an isotherm from D to C to B; in terms of tilt direction, this is reentrant behavior, although in fact B and D are distinct phases. For tetracosanoic acid such an isotherm would be at around room temperature. It can also be seen that many phase boundaries converge in this region; the B/C/D and C/A/A' triple points are separated by only about 3°C. Thus small temperature drifts and non-uniformity (Schlossmann et al. report $\pm 1^{\circ}$C) would give rise to time-dependence and 'bizarre coexistence' of structures.

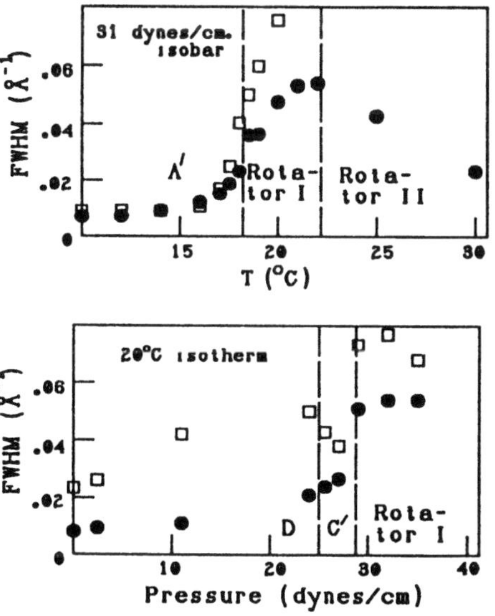

FIG. 2: Horizontal-plane widths of Peak I (●) and Peak II (□), the two first-order diffraction peaks, as a function of temperature at 31 dynes/cm. (top), and as a function of pressure at 20°C (bottom). (The hexagonal Rotator-II phase has only one first-order peak.) The resolution function (FWHM 0.008$Å^{-1}$) has *not* been deconvoluted; the narrowest peaks in the above data are resolution-limited.

The behavior of the diffraction peak widths [6,7] is also of interest. The peaks are resolution limited (resolution~0.01$Å^{-1}$) in the A and B phases, and at the lower-temperature end of the A′ and C phases; at higher temperatures in these phases, and in the Rotator phases, the peak widths are broad. In the D phase, the in-plane peak is resolution-limited while the off-plane peak is broad, implying a 2D 'liquid crystal' structure. The observation of peaks that are not resolution-limited, coupled with observations of bond-orientational order in monolayers transferred to solid substrates, has been used to argue that some monolayer phases are hexatic (see e.g., Kenn et al., these proceedings, and Ref. 4). Unfortunately, since the monolayers on water are powders, no direct measurement of bond-orientational order is currently possible, and the assumption that the process of transfer to solid substrates perfectly preserves the monolayer structure is difficult to justify. Of course, when strategies for making single crystal monolayers on water are found, it may turn out that some of the phases are indeed hexatic.

The peak widths also show reentrant behavior. They are broadest in the Rotator-I phase; as can be seen in Fig. 2, the peak widths anomalously narrow when the temperature is increased or if the pressure is decreased [8]. In 1942, Copeland et al. [10] found that the monolayer viscosity had a minimum in what we now know is a separate phase; they labelled as 'superliquid' the region where the viscosity anomalously increases with temperature. Our X-ray data now show that this anomaly in the mechanical properties has a microscopic origin. The physical basis for this 'reentrant disordering' is not yet clear. Since hydrocarbon chain cross-sections are not circular, one model of the Rotator phases is that the chains rotate about their axes in Rotator-II, while in Rotator-I they oscillate, and in the lower-temperature phases the cross-sections are frozen. An alternative picture (not involving chain motions) is that the chain cross-sections have completely random orientations in Rotator-II, partially random orientations in Rotator-I, and are ordered at lower temperatures. Whether any of these models explains the anomalous behavior of the peak widths remains to be established.

This work was supported by the US Department of Energy under grant no. DE-FG02-84ER45125.

1. e.g. S. Ställberg-Stenhagen and E. Stenhagen, *Nature* 156, 239 (1945)
2. e.g. M. Lundquist, *Chem. Scripta* 1, 5 (1971); 1, 197 (1971)
3. A.M. Bibo and I.R. Peterson, *Adv. Materials* 2, 309 (1990)
4. A.M. Bibo, C.M. Knobler and I.R. Peterson, *J. Phys. Chem.*, in press
5. B. Lin, M.C. Shih, T.M. Bohanon, G.E. Ice and P. Dutta, *Phys. Rev. Lett.* 65, 191 (1990)
6. B. Lin, Ph.D. thesis, Northwestern University, 1990; P. Dutta, in "Phase Transitions in Surface Films", edited by H. Taub, G. Torzo, H. Lauter and S. Fain (Plenum Publishing Co., 1991)
7. R.M. Kenn, C. Böhm, A.M. Bibo, I.R. Peterson, H. Möhwald, K. Kjær and J. Als-Nielsen, *J. Phys. Chem.*, 95, 2092 (1991)
8. M.C. Shih, T.M. Bohanon, J.M. Mikrut, P. Zschack and P. Dutta, submitted to *Phys. Rev. Lett.*
9. M.L. Schlossman, D.K. Schwartz, P.S. Pershan, E.H. Kawamoto, G.J. Kellog and S. Lee, *Phys. Rev. Lett.* 66, 1599 (1991)
10. L.E. Copeland, W.D. Harkins and G.E. Boyd, *J. Chem. Phys.* 10, 357 (1942)

Protein Recognition Processes at Functionalized Lipid Surfaces: A Neutron Reflectivity Study

D. Vaknin[1], *J. Als-Nielsen*[1], *M. Piepenstock*[2], *and M. Lösche*[2]

[1]Physics Department, Risø National Laboratory, DK-4000 Roskilde, Denmark
[2]Physical Chemistry, Mainz University, W-6500 Mainz, Fed. Rep. of Germany

Abstract. The specific binding of proteins to functionalized monolayers on aqueous subphases has been characterized by neutron reflectivity measurements. As a model for the investigation of a recognition process on a molecular length scale, streptavidin (SA) and biotin were chosen because of the high specific affinity between them. Reflectivities from the aqueous ($NaCl/H_2O$ or $NaCl/D_2O$) surfaces covered with the biotin-lipid monolayers before and after the adsorption of proteins were collected with a novel, fixed wavelength liquid surface neutron reflectometer. In quantitative terms, binding was found to occur at a biotin surface concentration as low as 1 molecule/1250 Å^2 (compare to $\sim$ 1 molecule/40 Å^2 for a densely packed monolayer). The reflectivity data are consistent with the formation of a monomolecular layer of protein at the interface, directly underneath the lipid monolayer. The thickness of the protein film is 44 ± 2 Å, comparable in size to the the smallest axis of the unit cell of 3-dimensional SA crystals (48 Å).

1. Introduction

Streptavidin is a protein isolated from *Streptomyces avidinii* and has an exceptionally high affinity to biotin with a stability comparable to that of covalent bonds [1]. Recent fluorescence microscopy studies have shown that streptavidin forms homogeneous domains underneath an aqueous surface covered with a lipid monolayer that includes biotin in its head group [2]. Electron diffraction from such systems, after transfer onto a solid support, revealed that the protein is in a crystalline form [3]. We have employed the neutron reflectivity technique to study these interfaces *in situ* on a molecular length scale.

2. Results and Analysis

Samples were prepared and measured on a Wilhelmy film balance situated on the sample table of a novel liquid surface reflectometer [4]

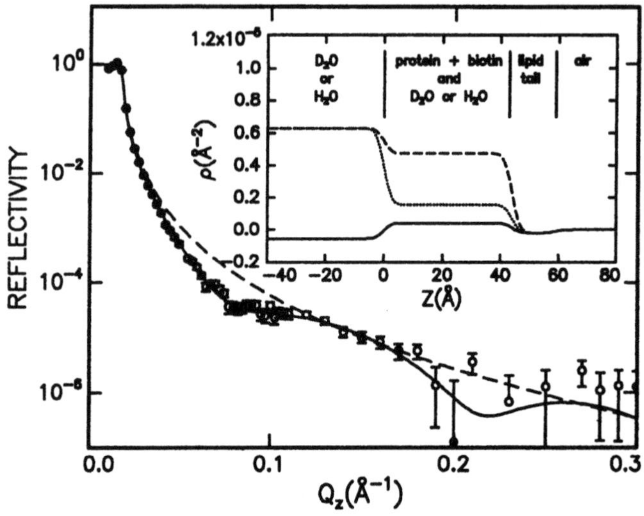

Figure 1: Measured and calculated neutron reflectivity curves from a D_2O solution of SA underneath a functionalized biotin lipid monolayer. The inset depicts the scattering length densities extracted from the reflectivity data (see text for details).

located at the guide hall of the cold source at Risø National Laboratory. The apparatus is a fixed wavelength ($\lambda \sim 4.8Å$) reflectometer operating similarly to an X-ray reflectometer built by Als-Nielsen and Pershan [5]. The neutron reflectivity from a SA solution in 0.5 M NaCl/D_2O underneath a monolayer of biotin lipid is shown in Fig. 1. A similar data set with H_2O as subphase was also taken. Included as dashed and solid lines, respectively, are the calculated reflectivities from the D_2O solution and from a model of the interface constructed as described below. In modeling the data, we have assumed that the interface consisted of two boxes of homogeneous scattering length densities. The first box consisted of the hydrocarbon tails of the biotin lipids, whereas the second contained the protein, the biotin head groups and interpenetrating water molecules. The best fit models describing the reflectivities taken with the H_2O and D_2O subphases are depicted schematically as an inset in Fig. 1. Two different scattering length density profiles corresponding to the reflectivity over D_2O as a subphase fit the data shown in Fig. 1 equally well. These two models, however, lead to different conclusions regarding the inplane packing of the protein monolayer. In the first model (dotted line, inset, Fig. 1) the calculated volume per protein suggests that the SA is closely packed, whereas in the second model (dashed line) the protein occupies only about 20% of the unit cell area.

Similar refinements of the reflectivity from the H_2O as a subphase lead to a different conclusion regarding the inplane packing of the proteins.

To resolve this problem we adopted the same approach of simultaneous refinement of the D_2O and H_2O data previously used by us in the structural study of a Langmuir monolayer [4]. The combined refinement is consistent with tight protein packing. Further support for this model comes from independent fluorescence microscopy observations on identically prepared samples which show pronounced excitation anisotropy from large 2D domains of fluorescently labeled SA, indicating that the protein packing occurs under severe geometric constraints. The protein layer thickness, $L_{SA} = 44 \pm 2$ Å, and average area per protein molecule in the film, $A = 2860 \pm 200$ Å^2 indicate intensive lateral contacts between the protein molecules that presumably lead to the formation of 2D crystals [3].

Further neutron reflectivity of isotopically substituted systems and X-ray reflectivity and GID measurements are under way for the purpose of working out the physical properties of these 2D systems.

Acknowledgement: We thank Prof. H. Ringsdorf for supplying the biotin lipid.

3. References

[1] N. M. Green, *Adv. Protein Chem.* **29**, 85 (1975).

[2] M. Ahlers, R. Blankenburg, D. W. Grainger, P. Meller, H. Ringsdorf and C. Salesse, *Thin Solid Films* **180**, 93 (1989).

[3] S. A. Darst, M. Ahlers, P. H. Meller, E. W. Kubalek, R. Blankenburg, H. O. Ribi and H. Ringsdorf, and R. D. Kornberg, *Biophys. J* **59**, 387 (1991).

[4] D. Vaknin, K. Kjaer, J. Als-Nielsen and M. Lösche, *Biophys. J.*, **59**, 1325 (1991).

[5] J. Als-Nielsen and P. Pershan, *Nucl. Inst. Meth.* **208**, 545 (1983).

Neutron Reflection from Liquid/Liquid Interfaces

T. Cosgrove[1], *A. Eaglesham*[2], *D. Horne*[3], *J.S. Phipps*[1], *and R.M. Richardson*[1,x]

[1]School of Chemistry, University of Bristol, Cantock's Close, Bristol BS8 1TS, UK
[2]Rutherford Appleton Laboratory, Neutron Division, Chilton, Didcot, Oxon. OX11 0QX, UK
[3]Hannah Research Institute, Physical Chemistry Group, Ayr, KA6 5HL, UK

Abstract. An experimental method for the measurement of neutron reflection from liquid/liquid interfaces is described. The strengths and limitations of the technique are described and its application to the study of adsorbed layers is discussed. Results are given for the protein β-casein and a block copolymer of polydimethylsiloxane-polyvinylpyridine oxide.

1. Introduction

This paper describes some experiments to measure the structure of adsorbed films at liquid/liquid interfaces using neutron specular reflection. Reflection is already used very successfully for studying liquid/air surfaces [1]. However, the importance of liquid/liquid interfaces in topics such as emulsion stability and detergency has lead us to develop the technique for studying them. No common liquid is sufficiently transparent to neutrons to make transmission of the neutron beam through (typically 10 cm) of the upper phase as shown in figure 1a a successful approach with currently available reflectometers. The method adopted in this work has therefore been to form a thin film (typically 10 µm) of the upper liquid and bring the beam through it as shown in figure 1b. These experimental considerations have imposed some restrictions on the applicability of the method, but it is possible to make useful measurements of adsorbed amount, film thickness and structure. Section 2 below describes the experimental method for maintaining a thin oil film over an aqueous subphase. Section 3 introduces the background theory relevant to reflection from "submerged" interfaces and section 4 outlines our approach to interpreting this reflectivity

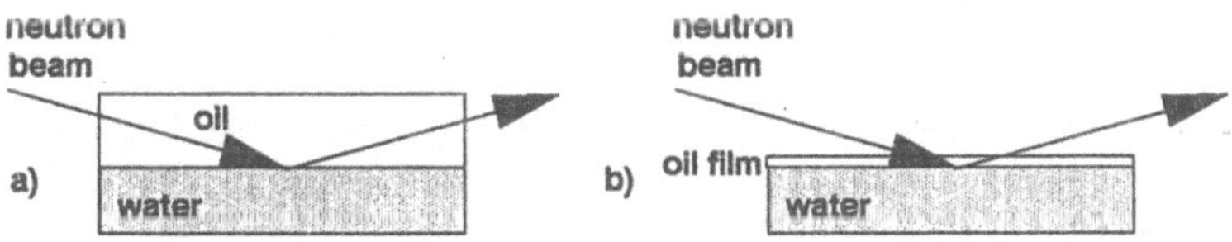

Figure 1 Schematic diagram of reflection geometry.

x author for correspondence.

Springer Proceedings in Physics, Vol. 61 Surface X-Ray and Neutron Scattering
Eds.: H. Zabel and I.K. Robinson

data. One of the great strengths of neutron scattering is the ability to adjust the scattering length density of materials by isotopic substitution, particularly deuterium for hydrogen. Section 5 discusses the choice of scattering length density of the bulk liquid phases and the adsorbate. The remaining sections describe some examples of neutron reflection studies of liquid/liquid interfaces with adsorbed block copolymer and protein.

2. Experimental method

Flat macroscopic oil/water interfaces have been prepared for neutron reflection experiments using the following procedure [4]. First a proud meniscus of the aqueous phase is formed in a PTFE trough. Some of the oil phase is then added as a thin film on the surface. The drainage of the film is surprisingly slow and would take several hours for the last few microns to go. The trough is contained in a hermetically sealed box (with fused silica windows for the neutron beam). The balance between condensation and evaporation of the oil film is controlled by varying the temperature of the liquid in the trough relative to the walls of the box. The thickness of the oil film could be deduced from the attenuation of the neutron beam and it proved possible to maintain a constant film thickness over a period of many hours while neutron reflectivity measurements were performed. In this work, hexane (rather than higher alkanes) has been used for the oil phase because it has a positive spreading coefficient on water which leads to a stable film. It has also been found preferable to study adsorbed films with an appreciable solubility in water so that the equilibrium of the adsorbed layer is not disturbed by the replacement of hexane solution by pure hexane condensate. Fortunately there are many interesting systems which satisfy these two criteria. A third criterion is that the adsorbate should not be appreciably attracted to the upper (i.e. hexane/air) interface. This is easy to satisfy because hexane/air is a low energy interface and surface tension measurements have indicated that the materials studied here are not appreciably adsorbed at it.

3. Theory of neutron reflection from submerged interfaces

In principle, the neutron reflection from an oil film on an aqueous subphase, with an adsorbed layer at the interface is given by the formula

$$R(Q)=R_1(Q)+\frac{(1-R_1(Q))^2 A(Q).R_2(Q)}{1-R_1(Q).R_2(Q).A(Q)}\ ,$$

where $R_1(Q)$ is the reflectivity of the air/oil interface, $R_2(Q)$ is the reflectivity of the oil/water interface and Q is the usual scattering vector. A(Q) is the transmission of the oil film:

$$A = \exp(-2\beta t/\sin\theta_1)$$

where β is the attenuation coefficient and t is the thickness of the oil layer. This formula assumes that oil film is thick enough to eliminate any interference of the rays scattered from the two interfaces. θ_1 is the angle of incidence of the neutron beam in the oil film and is determined by the angle of incidence in air (θ_0) and the change in neutron refractive index at the air/oil interface. For the purposes of modelling the measured reflectivity profile, these calculation have been done exactly, but it is useful to consider the approximate formula:

$$R(Q) \approx R_1(Q) + A(Q)\, R_2(Q) \ .$$

In the special case of an oil layer contrast matched to air the R_1 becomes zero and A becomes independent of Q:

$$R(Q) \approx A.R_2(Q) \ .$$

Since the objective of the measurements is to determine $R_2(Q)$, it is attractive to contrast match the oil layer to air (i.e. "null" oil) so that A is a "Q independent" scaling factor. If the aqueous subphase is (at least partly) D_2O, the value of A can be determined by the reflected intensity below the critical angle where $R_2 = 1$. However, to determine the distribution of the adsorbed material, it is generally necessary to measure the reflectivity from several combinations of scattering length density of the two liquids and the adsorbate as discussed in section 5.

4. Data interpretation and analysis

We have found it very useful to use Guinier type plots of the low Q part of the reflectivity data to obtain estimates of the layer thickness. It has been shown [2,3] that well above the critical angle the reflectivity is approximate to

$$R(Q) \approx \frac{16\pi^2}{Q^2} m^2 \exp(-Q^2\sigma^2) - \frac{32\pi^2}{Q^2} \Delta\rho <z> m \ \exp\left(\frac{-Q^2<z^3>}{6<z>}\right) + \frac{16\pi^2}{Q^4}\Delta\rho^2 \ ,$$

where $\Delta\rho$ is the contrast in scattering length density between the two phases, m is proportional to the amount adsorbed at the interface, $<z^n>$ is the "nth" power of distance from the interface averaged over the adsorbate distribution and σ is the second moment ($<z^2>-<z>^2$) of the adsorbate distribution ($\sigma^2 \approx <z^3>/(6<z>)$). The last term is the reflectivity that would be observed from an ideal, clean interface and is known as the Fresnel term (R_F).

If the two phases are contrast matched, only the first term contributes and a Guinier type plot of $\ln(R(Q).Q^2)$ vs. Q^2 has a slope of $-\sigma^2$ so the second moment of the adsorbate distribution is determined. If they are not contrast matched, the "Fresnel" term must be subtracted and the slope will give a number that is likely to be a close approximation to $-\sigma^2$.

A more detailed analysis has been done by simultaneously fitting layer models (calculated by the Abeles method [3]) to the data from all the combinations of contrast. The volume fractions (ϕ) distribution of the adsorbate is assumed to be the same and the scattering length density ρ is calculated for each contrast using the scattering length density of the pure material (ρ_p) and the solvent (ρ_s),

$$\rho = \phi\rho_p + (1-\phi)\rho_s \ .$$

5. Choice of liquid phase contrasts

The combinations of scattering length densities chosen for the liquid phases depends upon the availability of deuterated versions of the adsorbate. If the adsorbate is only available as a hydrogenous material it is likely that its scattering length density will not contrast significantly with the null oil or water. In this case, three combinations of contrast have been used where possible.
i) Null oil/D_2O allows R_2 to be measured on an absolute scale and is sensitive to the amount of penetration of the adsorbate into the D_2O. It will not be sensitive to adsorbate in the null oil phase.
ii) Null oil/null water gives an unambiguous measure of the adsorbed layer thickness (by a Guinier plot) but the reflectivity may be too weak to measure satisfactorily.
iii) Deuterated hexane/D_2O should give the best contrast with a hydrogenous adsorbate. Unfortunately the observed reflectivity is a combination of the air/hexane and hexane/water interfaces. If the hexane is a mixture of C_6H_{14} and C_6D_{14} to give a scattering length density less than D_2O then two critical angles are observed and the step between them can be used to determine A(Q).

6. Neutron reflection results from β-casein

The milk protein, β-casein, is an extremely good emulsifier and neutron reflection measurements were made to determine its distribution at a water/hexane interface as part of a broader study of its properties. Figure 2 shows Guinier type plots of β-casein adsorbed from 5 x 10^{-3} m solution. The upper phase is air contrast matched hexane (actually $C_6H_{13}D$) and the lower is

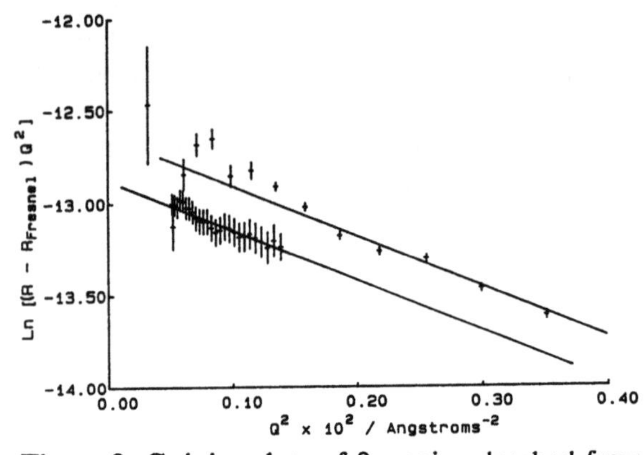

Figure 2 Guinier plots of β-casein adsorbed from a) D_2O and b) null water.

either null water or D_2O. The similarity in the second moments from the two plots suggest $\lambda(\sigma = 17Å)$ that the protein is predominantly in the aqueous phase since the D_2O data will be relatively insensitive to any protein in the hexane film. A simultaneous fit of a two layer model to both contrasts gives a primary layer at the D_2O surface which is about 15Å thick with a volume fraction of 0.6. The secondary layer is 50Å thick and has a much lower protein volume fraction of 0.15. This suggests that some parts of the molecule are pinned to the surface while more hydrophilic residues penetrate deep into the water.

7. Neutron reflection results from polydimethylsiloxane-polyvinylpyridine diblock copolymer

This material (M_n = 9000, M_w/M_n = 1.2, 52% of the monomers are PVPO) is an example of an amphiphilic molecule. The PDMS block is only soluble in the hexane while the PVPO block is water soluble. Since it was anticipated that some of this molecule would form a layer in the hexane phase, a deuterated layer was included in the contrasts measured.

The reflectivity of the PDMS PVPO adsorbed from 0.2 g dm^{-3} solution in D_2O at a null hexane is shown in figure 3. Analysis shows that the presence of the polymer does not make the hexane/water interface diffuse and that there is a 25 Å layer of polymer with volume fraction in D_2O of between 0.1 and 0.2 depending on the diffuseness of the junction between the polymer and bulk water. This data is also consistent with a polymer layer of similar thickness in the hexane since this would not change the form of the autocorrelation function so we can conclude that $\phi(\text{in } D_2O) \leq 0.25$. Fitting a single layer model to the null oil/null water contrast shows a 42 Å layer but since the scale factor is not known strong conclusions about the polymer volume fraction cannot be drawn.

With the above results to guide us, we have sought a polymer volume fraction profile that fits all the combinations of liquid phase scattering length density.

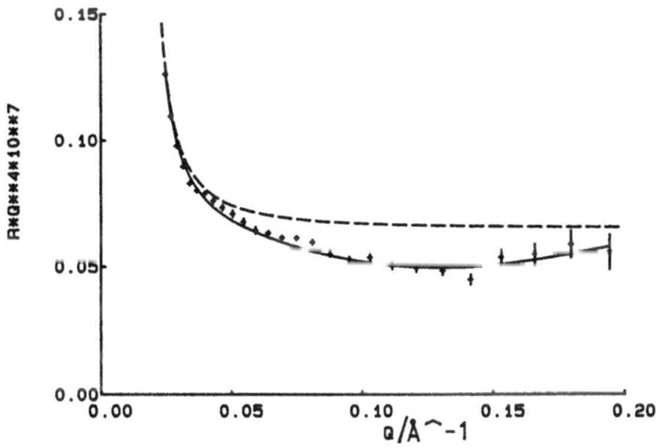

Figure 3 Model fit to reflectivity from PDMS-PVPO adsorbed at null hexane from D_2O. The dashed line would be observed for an ideal, clean interface.

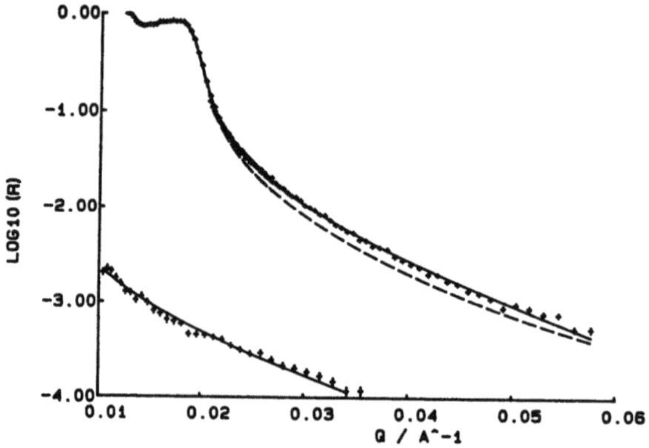

Figure 4 Model fit to reflectivity PDMS-PVPO at null hexane/null water and 60% deuterated hexane/D_2O. The dashed line would be observed from the latter if no polymer were present.

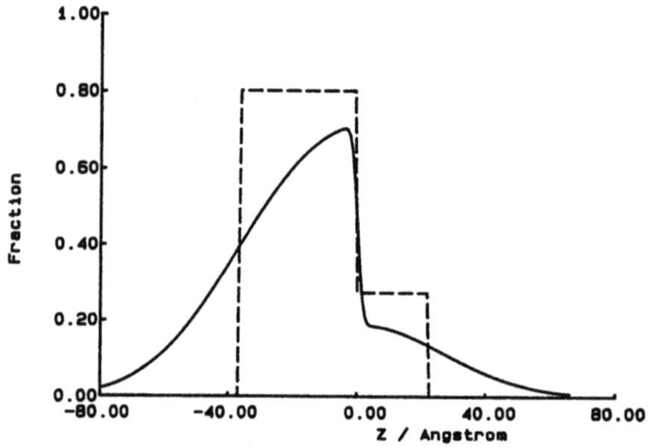

Figure 5 Volume fraction profile for PDMS/PVPO at hexane water interface deduced from neutron reflectivity measurements. The dashed line indicates the profile before the interfacial diffuseness is introduced.

The main problem was to find a profile which would only weakly modify the reflectivity from null hexane/D_2O but strongly modify that from the deuterated hexane/D_2O. This suggested a substantial polymer layer (i.e. high ϕ) in the hexane but a rather tenuous one in the water. We first considered the obvious model of equal amounts of PDMS in the hexane and PVPO in water. This was not able to fit d-hexane/D_2O and the null hexane/null water data without invoking an impossibly high scaling factor for the null/null data. Single layers of a mixture of PDMS and PVPO were then considered either in the water phase or the hexane phase. Neither could account for the measured reflectivities. The next step was to use one layer of mixed polymer in the hexane and a second layer in the water. This was able to account for all three data sets and two are shown in figure 4. A 35 Å layer of volume fraction 0.8 is in the hexane and a 22 Å layer of volume fraction 0.2 in the water. Both the junctions between polymer layers and bulk liquid were diffuse and were characterized by error function profiles with standard deviation 26 Å was shown in figure 5. Apparently most of the polymer resides in a polymer rich layer in the hexane and a few strands (presumably PVPO) are penetrating into the water phase. The insolubility of PDMS in water is perhaps causing the PVPO to be withdrawn from the water phase.

8. Conclusion

We have shown that neutron reflection can be used to determine the distribution of adsorbate molecules at liquid/liquid interfaces. The method used has some limitations to its applicability but should allow a number of systems to be studied in detail. In particular, it is not necessary to prepare deuterated versions of the adsorbing species since the use of a deuterated top film is possible.

Acknowledgements

We wish to thank Dr. Jeff Penfold and the late Dr. John Herdman for help in carrying out these experiments on the CRISP reflectometer at the Rutherford Appleton Laboratory and Dr. Simon Biggs for use of the copolymer. One of us (JSP) is grateful for financial support from SERC and Shell Research.

References

1. R.M. Richardson and S.J. Roser, Langmuir, 1991 (in press).
2. T.L. Crowley, D.Phil.Thesis, Oxford University, 1984.
3. J. Penfold and R.K. Thomas, J.Phys.Condens.Matter 2 (1990) 1369.
4. T. Cosgrove, J.S. Phipps and R.M. Richardson, submitted to Colloids and Surfaces.

Polymer Interfaces Analysed on a Nanometer Scale: X-Ray and Neutron Reflectometry

M. Stamm

Max-Planck-Institut für Polymerforschung,
Postfach 3148, W-6500 Mainz, Fed. Rep. of Germany

Recently the techniques of X–ray and neutron reflectometry (XR, NR) have also been applied to polymer films and interfaces. While the use of X–rays in polymeric systems is largely limited to specific cases because of a general lack of contrast between most polymers, NR can be widely applied with deuterated polymers. While the deuteration might cause for instance surface enrichment processes on itself, it can be nicely used in many cases to generate a large contrast between components. Since the resolution of the techniques is the sub–nanometer range, special care has to be taken to prepare very smooth and homogeneous polymer films. It then is possible to investigate various aspects of polymer interfaces on a nanometer scale. Several examples are discussed (see [1,2] for a general review).

With XR for instance the evolution of a small interface between two largely incompatible polymers can be investigated as a function of annealing temperature [3]. The difference in the glass transition between components largely influences the interface width at low temperatures. Also the movement of the interface during the interdiffusion of different polymers may be followed accurately by XR when a specific marker is introduced at the interface [4,5]. At very initial stages of interdiffusion polymer specific deviations from a $\sqrt{t}$ – behaviour are observed. Similarly with NR various time regimes of interfacial broadening are observed when protonated and deuterated polymers start intermixing in the melt [6,7]. This peculiar behaviour is consistent with common pictures of polymer diffusion, but may be influenced by the specific chain conformation of polymers at the interface. While most polymers have a molecular size of several nanometers, also segmental movement of chains at the interface can be followed by XR and NR due to the excellent resolution of those techniques. For a unique interpretation of the data other supplementary techniques may be used [2,8] like the nuclear reaction analysis technique which yield direct information on hydrogen [9] or deuterium [10] profiles.

[1] T.P. Russell, Mat. Sci. Rep. 5 (1990) 173

[2] M. Stamm, Adv. Pol. Sci. 100 (1991) in press

[3] S. Hüttenbach, M. Stamm, G. Reiter, M. Foster, Langmuir, in press

[4] P.F. Green, C.J. Palmstrom, J.W. Mayer, E.J. Kramer, Macromolecules 18 (1985) 501

[5] G. Reiter, S. Hüttenbach, M. Foster, M. Stamm, Macromolecules 24 (1991) 1179

[6] A. Karim, A. Mansour, G.P. Felcher, Phys. Rev. B 42 (1990) 6846

[7] M. Stamm, S. Hüttenbach, G. Reiter, T. Springer, Europhys. Lett. 14 (1991) 451

[8] P.F. Green, B.L. Doyle, in "New Characterisation Techniques for Thin Polymer Films", H.M. Tong, L.T. Nguyen (eds.), J. Wiley, New York (1990), p. 139

[9] D. Kuhn, F. Rauch, A. Götzelmann, G. Reiter, M. Stamm, to be published

[10] U.K. Chaturvedi, U. Steiner, O. Zak, G. Krausch, J. Klein, Phys. Rev. Lett. 63 (1989) 616

Neutron Reflection from Polymers Adsorbed at the Solid/Liquid Interface

*T. Cosgrove, J.S. Phipps, and R.M. Richardson**

School of Chemistry, University of Bristol,
Cantock's Close, Bristol BS8 1TS, UK

Abstract. Neutron reflection measurements have been made on block copolymers adsorbed onto single crystal quartz. The effect of solvent quality on the segment density distribution of the adsorbed polymer layer has been clearly observed. The presence of a layer of bound water at the quartz surface has also been observed, emphasising the need for caution when interpreting the results of this type of experiment.

1. Introduction

There is much current interest, both experimental and theoretical, in the segment density distributions of adsorbed copolymers, as a result of their commercial importance as stabilisers of colloidal dispersions. Recently several experimenters have been using the specular reflection of neutrons to investigate such adsorbed layers from a range of solvents. The most commonly used solid substrate is amorphous or single crystal quartz, since it is possible to pass a beam of neutrons through a large (~10cm in length) sample and study the solid/liquid interface directly. We have used the technique to study the adsorption of AB block copolymers of polystyrene / poly(2-vinylpyridine) (a-PS/b-PVP) and polydimethylsiloxane / polyhydrogenmethylsiloxane (a-PDMS/b-PHMS).

2. Sample preparation

Figure 1 shows the X- ray reflectivity profile of a cleaned quartz single crystal in air. The broad interference fringe visible in the data can be fitted with a single 17 Å layer of lower scattering length density than the quartz. By heating the sample under vacuum to about 150°C and transferring it to a sealed cell under dry nitrogen the fringe may be removed leaving an apparently clean surface. When water is allowed into the cell the fringe gradually returns over several hours. A similar experiment using D_2O and neutron reflectivity suggests that the layer consists of water penetrating into the quartz surface.

As a result of the above observations all samples for polymer adsorption studies were prepared by first heating the quartz crystal under vacuum, and adsorbing the polymers under dry conditions. For the a-PDMS/b-PHMS sample a catalyst (dibutyl tin diacetate) was used to chemisorb the PHMS block to the

* Author for correspondence

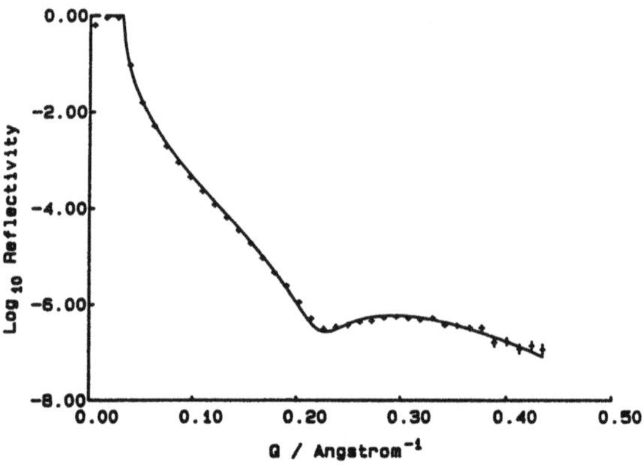

Figure 1. X- ray reflectivity of a clean quartz crystal.

surface. Each sample was then attached to a small teflon trough containing solvent. After the adsorption stage high purity solvents, both hydrogenous and deuterated, were used as supplied.

The adsorption of block copolymers is essentially an irreversible process [1], so the reflection experiments were carried out using pure solvents rather than polymer solutions. This approach allows the structure of the adsorbed layer to be examined under different solvency conditions with a constant adsorbed amount. The desired solvent contrasts were achieved by emptying the cell and flushing several times with solution of the correct contrast before filling. Details of the polymers used are given in Table I.

Table I

Polymer	M_n (a Block)	M_n (b Block)	M_w/M_n
a-PS/b-PVP	8000	4000	1.2
a-PDMS/b-PHMS	50000	700	1.12

3. Neutron Reflectivity Measurements

Figure 2 shows the reflectivity multiplied by Q^4 of the a-PS/b-PVP copolymer under various solvency conditions, in each case with the solvent contrast-matched with the quartz block. At room temperature toluene is a good solvent for the PS block, cyclohexane is slightly worse than a theta solvent and octane is a non-solvent. All are non-solvents for the PVP block. A change in the width of the interference fringe resulting from the polymer layer as the solvent quality for the PS block decreases is clearly visible. Figure 3a shows the polymer volume fraction profiles that correspond to the solid line fits to the data in Figure 2. There is a region of high volume fraction within the first 50Å of the

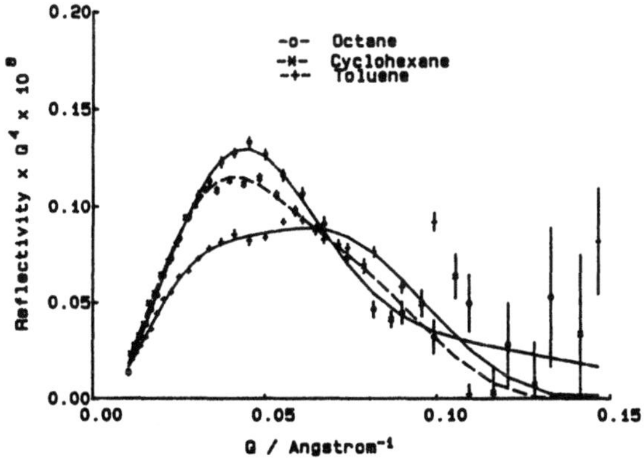

Figure 2. Reflectivity x Q^4 of adsorbed a-PS/b-PVP copolymer under differing solvency conditions

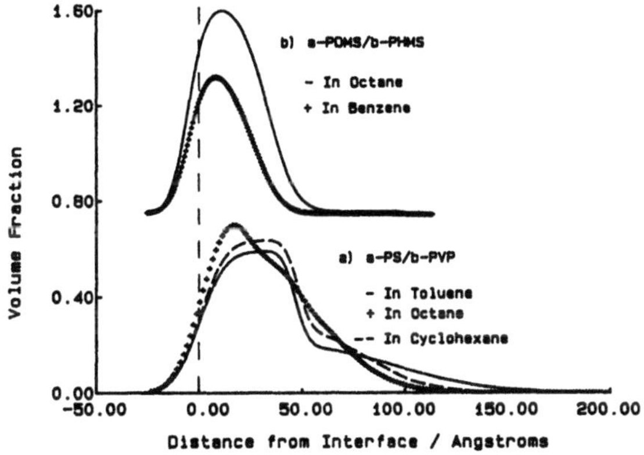

Figure 3. Polymer volume fraction profiles obtained from fitting the reflectivity data. The vertical dotted line represents the mean position of the quartz surface.

surface; this presumably contains the large, adsorbing PVP block. In toluene there then appears to be a sharp decrease in the volume fraction and a 'tail' region of slowly decaying density over the next 150Å. Since the radius of gyration of the PS block is estimated to be 25Å it appears to be quite stretched. As the solvent quality for the PS block decreases so the tail region contracts towards the surface. The profiles were generated by modelling the adsorbing block as a single layer and the non-adsorbing block as a half-gaussian decay, and allowing the layer thickness, the decay length, the volume fractions and the interfacial roughnesses to vary during the fitting. The surface roughness of the quartz was fixed at 10Å in accordance with the x-ray measurement on the dry quartz. The same basic profile shapes were achieved when just a two layer model was used; in this case the optimum fits occur when the interfacial roughness between the second layer and the solvent becomes very large, which is equivalent to the second layer being a gradual decay rather than a well-defined block. However, a fairly sharp interface between the two layers is necessary to fit the interference fringe pattern.

The experiment was repeated with toluene as the solvent, taking measurements at two solvent contrasts: contrast-matched and deuterated toluene respectively. There is good agreement between the two measurements at contrast-match, but the measurement with deuterated toluene will only agree with the contrast-match if the block adjacent to the surface is split into two regions of different density. The first 20Å corresponds to the water layer observed with x-rays.

Figure 3b shows the apparent volume fraction profiles of the copolymer of a-PDMS/b-PHMS obtained from neutron reflectivity measurements using octane (good solvent for PDMS) and benzene (slightly better than theta). Each of these profiles was measured at two different contrasts, with the contrast-matched and deuterated solvents respectively. The data was fitted to a single layer, with the interfacial roughness with the quartz again fixed at 10Å and the roughness with the solution as a variable parameter. In each case the volume fraction at the surface appears to be very high, although some desorption on changing the solvent from octane to benzene is apparent. It thus appears that the polymer was physically, rather than chemically adsorbed. Recent results [2] have shown that for adsorbed polystyrene/poly(ethylene oxide) copolymers even trace amounts of water in the solvent may be sufficient to displace an adsorbed polymer. Changing the solvent to methanol, which can hydrogen-bond with the surface, resulted in the complete desorption of the polymer, since the reflectivity with the contrast-matched solvent was reduced to virtually zero.

For both the benzene and the octane nearly all of the polymer appears to be within the first 50Å of the surface, which is surprising given the high initial volume fraction, and the estimated radius of gyration of 60Å. However, it is possible that there is some bound water contained in this region, and so the adsorbed amount is not as high as is first apparent.

Conclusions

Neutron reflectivity has been shown to be sensitive to the segment density distribution of adsorbed polymers at the quartz/water interface, and the effect of solvent quality on the distribution has been observed. The results can only be successfully interpreted by inclusion of a layer, corresponding to the bound water observed with x-rays, at the surface of the quartz.

Acknowledgements

We wish to thank Dr.Jeff Penfold and the late Dr. John Herdman for their help in carrying out the experiments with a-PS/b-PVP on the CRISP reflectometer at the Rutherford Appleton Laboratory, and Dr. Pierre Terech and Dr. Francois Rieutord for their help with the D17 spectrometer at ILL for the experiments with a-PDMS/b-PHMS. We thank also Dr.T.Heath for the synthesis of the a-PS/b-PVP copolymer. One of us (JSP) is grateful for financial support from SERC and Shell Research, Thornton, UK.

References

1. M.V.Tirell, S.S.Patel, Ann.Rev.Phys.Chem. 14, 597 (1989)
2. M.Hair, C.Tripp, unpublished results

Part V

Electrochemistry

Electrochemical Roughening of Au(110) Single Crystal Electrodes

K.M. Robinson[1], *W.E. O'Grady*[1], *and I.K. Robinson*[2]

[1]Code 6170, Chemistry Division, Naval Research Laboratory, Washington, DC 20375, USA
[2]AT&T Bell Laboratories, Murray Hill, NJ 07974, USA

Abstract. The surface roughness of a Au(110) electrode was studied as a function of electrode potential. The electrode potential was cycled between the double layer region and the oxide and hydrogen regions. Measurements of the crystal truncation rod passing through the (111) reflection were made ex-situ. The results reveal that the surface roughness is not significantly changed by cycling the potential between the double layer and the oxide regions, (.3V - 1.4V vs. Pd/H_2, 0.1M $HClO_4$). The roughness is decreased when the potential is cycled into the hydrogen region, -0.4V.

1. Introduction

The surface structure of Au(110) single crystals has been frequently studied in vacuum [1-3]. The 1x2 reconstruction, which has been studied theoretically using embedded atom models [4,5], has repeatedly been observed by LEED [2] and x-ray diffraction [1]. The stability of the Au(110) surface was studied as a function of the electrochemical potential by LEED analysis [2]. It was reported that the Au(110) surface reconstruction was lifted by the oxidation of the surface, however these samples were vacuum prepared. In the work presented here, electropolished electrodes were examined as a function of potential. Recent work on flame-annealing of gold crystals has demonstrated that the roughness of these electropolished electrodes is sufficient to diminish the intensity of the scattering from the reconstruction [6]. Therefore, the roughness, and possible self-annealing, of the Au(110) electrode was studied.

The roughness of the Au(110) electrode was studied by measuring the crystal truncation rods, CTRs. These rods are described as lines of scattering perpendicular, in k-space, to the terminated surface passing through the bulk Bragg positions [7]. More precise details of CTRs are contained elsewhere [7,8]. A perfectly terminated flat surface would produce rods with a $1/\Delta q_z^2$ intensity dependence. Surfaces that are more rough deviate from the ideal dependence according to a roughness parameter, β, which is an estimate of the fractional occupancy of subsequent layers. The integrated intensities measured are fit to

$$|F(q_z)|^2 = N_1^2 N_2^2 \frac{(1-\beta)^2}{[1+\beta^2-2\beta\cos(q_z c)]} \frac{1}{4\sin^2(\frac{1}{2}q_z c)}$$

where N^2_i are the crystal dimensions in the surface plane and c is the real space unit cell dimension in the surface normal direction.

2. Experimental

The preparation of the Au(110) electrode is discussed elsewhere [6]. Briefly, the surface is electropolished in a HCl:ethanol:glycol mixture at 4.5V [9]. This repeatedly produces a surface with a roughness parameter, $\beta = 0.56 \pm 0.08$. Chloride ion contamination from the electropolishing was removed by placing the electrode in boiling concentrated HNO_3. The electrode was emersed in 0.1M $HClO_4$ (Ultrex grade) under potential control at 0.3V (vs. Pd/H_2). The electrode potential was cycled, 25mV/sec, between the emersion potential, 0.3V, and two extremes, 1) oxide region, 1.4V, and 2) hydrogen region.

The electrode was removed from solution, under potential control, and mounted on a four-circle diffractometer and aligned according to bulk diffraction indices. Rocking scan measurements were made along the (1,1,l) truncation rod. The scans were background subtracted and integrated according to standard procedures [8]. Once the scan of the rod was completed, the crystal was re-emersed, or re-electropolished depending upon the desired experiment. The ex-situ nature of this work prevented any detail concerning the structural changes possibly caused by OH adsorption or reduction of the oxide layers. For details concerning these regions, in-situ scattering is necessary to be more electrochemically correct.

3. Results and Discussion

The data and best fits for the electropolished surface and the surfaces in the oxide and hydrogen regions are shown in figure 1. The data are averages of 5-7 separate electropolished and cycled samples. The electropolishing repeatedly produces roughness parameters of $\beta = 0.56 \pm 0.08$. There is no statistical difference

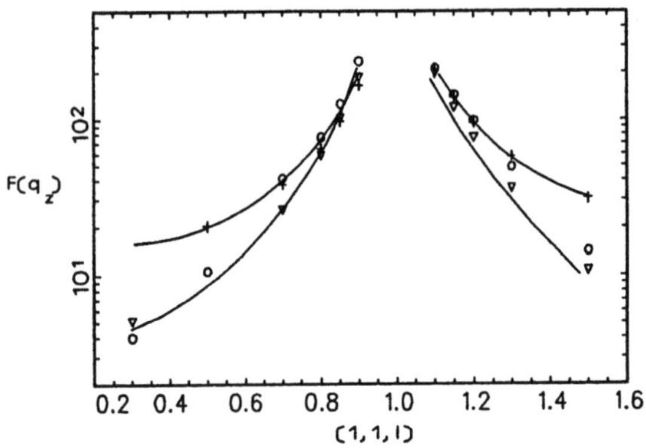

Figure 1: (1,1,l) truncation rod data for the electropolished surface (O), and the surfaces cycled into the oxide (∇) and hydrogen (+) regions. The solid lines are the best fit to the data.

between the electropolished roughness and the oxidized surface. This is in agreement with studies done on Au(100) crystals [6].

The data for the hydrogen region are fit with a roughness of $\beta = 0.27 \pm 0.09$. This decrease in the roughness supports observations in vacuum of self-annealing of Au crystals by repetitive cycling of the potential into the hydrogen production region [2,10]. There does appear to be a time-dependence to the smoothing of the surface, however after approximately 10 cycles (15 minutes), the roughness parameter remains constant so only data after 15 minutes were used in the fitting. There was no observation at any potential of the 1x2 reconstruction.

By assuming that upon reduction the gold atoms occupy (110) atomic positions, as opposed to a close-packed structure seen on Au(100), β can be related to an rms surface roughness by rms = $d_\perp \beta^{1/2}/(1-\beta)$ (for Au(110) $d_\perp$ = 1.44 Å). Michaelis etal. reported the reappearance of the 1x2 reconstruction after reduction of the oxide, however they did not cycle into the hydrogen production region [2]. The electropolished and oxidized samples have an rms = 2.5± 0.3 Å which is very close to the Au-Au distance or one lattice distance. The surface cycled into the hydrogen production region has an rms = 1.0± 0.4 Å which is very close to a half lattice distance. The rms values correspond to the presence, by electropolishing, and disappearance, by cycling the potential into the hydrogen region, of a missing row reconstruction, however the lack of any 1x2 surface scattering does not support this interpretation. One possible explanation would be small domains of 1x2 reconstructions, which are predicted by the embedded atom model [5]. These small domains would produce low intensity surface scattering on the order of the background noise. Future in-situ scattering work will aide in determining the driving mechanism for the self-annealing and the extent gold oxidation-reduction plays in the annealing.

This work is funded by the Office of Naval Research. KMR would like to acknowledge the support of the National Research Council Fellowship Program. In addition, we would like to acknowledge AT&T for the use of X16B and the NSLS which is funded by the Department of Energy.

References

[1] E. Vleig, I. K. Robinson and K. Kern, Surface Sci. 233 (1990) 248.
[2] R. Michaelis and D. M. Kolb, Surface Sci. Lttrs. 234 (1990) L281.
[3] E. van de Riet, H. Derks and W. Heiland, Surface Sci. 234 (1990) 53.
[4] L.D. Roelofs, S.M. Foiles, M.S. Daw and M.I. Baskes, Surface Sci. 234 (1990) 63.
[5] S.P. Chen and A. F. Voter, Surface Sci. Lttrs 244 (1991) L107.
[6] K.M. Robinson, I.K. Robinson and W.E. O'Grady, in preparation.
[7] I.K. Robinson, Phys.Rev. B 33 (1986) 3830.
[8] R. Feidenhans'l, Surface Science Reports 10 (1989) 105.
[9] J.L. Whitton and J.A. Davies, Steel Corrosion Mechanisms 111 (1964) 1347.
[10] D.M. Kolb and J. Schneider, Surface Sci. 162 (1985) 764.

Part VI

Thin Films and Multilayers

Reflectivity Studies of Thin Au Films and Au Bicrystals with Grain Boundaries

E. Burkel[1], *M. Fitzsimmons*[2], and *M. Müller-Stach*[1]

[1]Sektion Physik der Universität München, W-8000 München 22, Fed. Rep. of Germany
[2]LANSCE, Los Alamos National Laboratory, Los Alamos, USA

X-ray reflectivity is an ideal method to get information on buried interfaces and, therefore, well suited for the study of artificial grain boundaries in bicrystals. Such bicrystals with (001) twist grain boundaries are produced by hotpressing two thin single-crystal films, one of 25nm thickness and the other of 400nm thickness, in air at 523K for 1-2 hours. The single-crystals are epitaxially grown onto single-crystal NaCl substrates. Our reflectivity studies of the single-crystal films either still on the substrate or dissolved and mounted on a Si-wafer and the bicrystals themselves revealed in each case strong diffuse non-specular intensity. The existence of this non-specular intensity can be explained by lateral correlations of atoms which are detected because of an additional component of the scattering vector parallel to the surface. Contributions in the diffuse non-specular intensity originating from the artificial grain boundary could be identified by performing experiments with its in situ formation and destruction through annealing. The non-specular reflectivity is observed to be dependent on the twist angle of the grain boundary.

During the last years a variety of experiments has been performed on the analysis of structural properties of artificial grain boundaries in bicrystals [1,2,3]. It was found that the atomic density within the boundary volume, which can be also described as buried interface, is reduced in comparison to the bulk density [3]. Due to the sensitivity of reflectivity studies on the electron density, this method is particularly well suited to get new information about such bicrystals. It is also important that these studies allow a more stable mounting of the sample compared to conventional scattering experiments [1,2,3]. The Au-films and bicrystals - made by hotpressing two Au-films with thicknesses of 25nm and 400nm - were mounted on commercially polished Si wafers. For the interpretation of reflectivity studies of bicrystals it is very important to have information on its constituents, both single crystal films. Therefore, reflectivity studies of a 400nm and a 25nm thick crystal were performed. The measurements were done using the 3-axis spectrometer D4 at HASYLAB (DESY,FRG) with a wavelength of $\lambda = 1.11$Å and

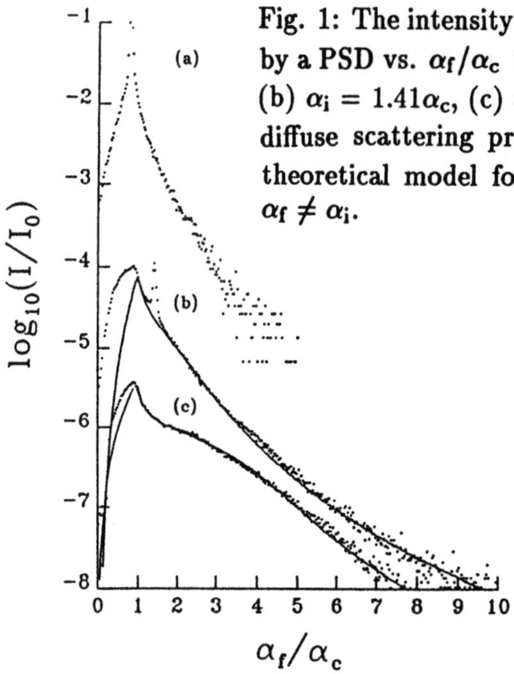

Fig. 1: The intensity of a thick Au sample recorded by a PSD vs. α_f/α_c for (a) $\alpha_i = 0.8\alpha_c$, (b) $\alpha_i = 1.41\alpha_c$, (c) $\alpha_i = 2.82\alpha_c$, with fitted diffuse scattering profiles calculated from a theoretical model for the region $\alpha_f > \alpha_c$ and $\alpha_f \neq \alpha_i$.

$\Delta\lambda/\lambda = 10^{-3}$ [4,5]. Reflected intensity from the specimen was measured by a position sensitive detector as a function of exit angle α_f. The result for the 400nm thick crystal is shown in Fig. 1 with the reflectivity profile vs. α_f/α_c for different α_i. The profiles are normalised to the integrated intensity in the case of $\alpha_i = 0.8\alpha_c$. The reflectivity of thick Au samples is sharply peaked about the specular beam, and shows an additional peak in the non-specular diffuse scattering around the critical angle α_c of Au. The diffuse scattering decays rapidly as the magnitude of the component of the scattering vector perpendicular to the surface increases, regardless whether or not the specular reflected beam can be observed.

An analysis of the non-specular reflectivity, which provides information about the height-height correlation function of the surface, $< z(0,0)z(X,Y) >$, can be accomplished by combining the theoretical treatments of surface scattering by Dietrich and Wagner [6] with that of Sinha et al. [7]. Using the distorted wave approximation, the intensity $I(\vec{R})$ scattered by a perturbation of a smooth surface is given as [6],

$$I(\vec{R}) = \left(\frac{r_e}{R}\right)^2 \left|(1 - \hat{R}\hat{R})\,\mathbf{T}(\alpha_f)\,\mathbf{T}(\alpha_i)\,\hat{e}\right|^2 I_0\,|F(\vec{r})|^2 \,. \tag{1}$$

In equation (1), $\mathbf{1}$ is the identity matrix, $\hat{R} = \vec{R}/R$, $\mathbf{T}$ is the Fresnel transmission matrix, $\hat{e}$ is the electric polarisation vector, I_0 is the incident beam intensity, and $F(\vec{\tau}) = \int \tilde{\rho}(\vec{r}) e^{i\vec{\tau}\cdot\vec{r}} d^3r$ is the structure factor of the perturbation. The observed intensity represents an intensity measurement of the fluctuations of surface heights averaged over the entire sample. Therefore $|F(\vec{\tau})|^2$ can be replaced by its ensemble average, $<|F(\vec{\tau})|^2>$. Assuming a self-affine rough surface with an effective cutoff length ξ for the roughness leads to the height-height correlation function $< z(0,0)z(X,Y) > := C(R) = \sigma^2 \exp(-(R/\xi)^{2h})$. The non specular diffuse component of this ensemble average can be calculated as [7],

$$< |F(\vec{\tau})|^2 >_{ns} = 2\pi S_0 \rho_0^2 \frac{e^{-\Re(\tau_\perp^2)\sigma^2}}{|\tau_\perp|^2} \int_0^\infty R(e^{|\tau_\perp|^2 C(R)} - 1) J_0(\tau_{||}, R) dR \, . \quad (2)$$

Here z(X,Y) is assumed to be a Gaussian random variable of variance $\sigma^2 = < z^2 >$, which is the saturation value for the mean-square roughness. The exponent $h\,(0 < h < 1)$ represents the texture of the roughness. For example, if a surface shows smooth hills, h would be close to 1, whereas extremely jagged surfaces will have a very small exponent h [7]. Eqn. (2) is valid whether or not $\tau_\perp$ has an imaginary component.

The solid lines in Fig. 1 are the best fits acchieved by the discussed model, using a non-linear least squares fitting routine. Only the part of the profiles above the critical edge were considered, because of possible geometrical effects on the low α_f data, that are probably associated with the instrument [8]. Average values of $\sigma = 4.5$Å and $h = 0.57$ are obtained from best fits at the intensity data for $\alpha_i \geq 2.82\alpha_c$ [4,5]. The rms fluctuation height σ of the surface is roughly the size of the unit cell of Au. The obtained value of h implies that the surface shows a random distribution of fluctuation heights.

In the case of a 25nm thin Au film the penetration depth for $\alpha_i = 2.82\alpha_c$ is about 4 times the sample thickness. So influence of the interface Si-Au on the diffuse reflectivity is to be expected. The observable modulation of the non-specular diffuse scattering is consistent with the film thickness of 25nm and is caused by an interference between X-rays reflected because of the discontinuity in the electron density at the upper (air-Au) and at the lower interface (Au-Si) of the crystal [4,5].

Non-specular diffuse intensity is also observed from bicrystals of Au that contain large area (001) twist grain boundaries of different twist angles Θ_t, as shown in Fig. 2. The three intensity profiles that correspond to bicrystals with sintered interfaces show pronounced differences of their diffuse scattering profiles. As the twist angle decreases, the diffuse scattering broadens. The sharp specular reflection from the bicrystal with $\Theta_t = 28.1°$ suggests that portions of its interface are not as rough as

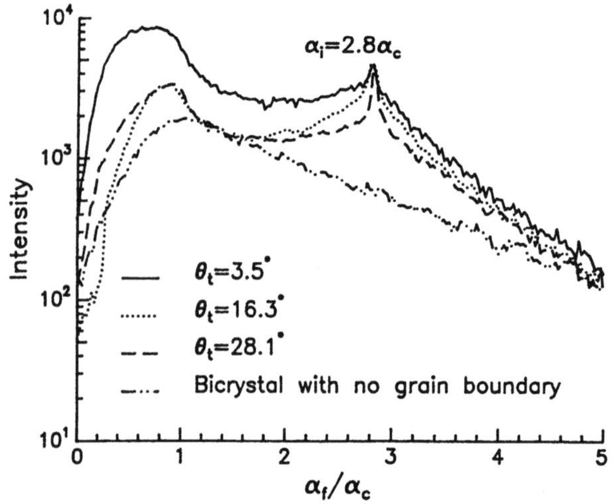

Fig. 2: The diffuse scattering for $\alpha_i = 2.82\alpha_c$ of a bicrystal with $\Theta_t = 3.5°$ (solid), $\Theta_t = 16.3°$ (dotted), $\Theta_t = 28.1°$ (dashed) and no grain boundary (– · ·).

the interface in the $\Theta_t = 3.5°$ bicrystal, where no sharp specular beam is observed. In the case of a specimen that was formed by just lying a thin Au film on top of a thick one, thus containing no large area grain boundary, the non-specular reflectivity is drastically reduced, showing that grain boundaries produce additional diffuse scattering.

For a further prove that parts of the diffuse scattering shown in Fig. 2 originate from the interface between the Au crystals, measurements of a bicrystal ($\Theta_t = 28.1°$) as a function of temperature were made. Reflectivity profiles taken at temperatures of 295K and 428K for $\alpha_i = 2.82\alpha_c$ are shown in Fig. 3(a) and 3(b), respectively. After annealing at 643K for 30minutes the reflectivity profile from the same bicrystal was recorded again. It is shown as the dashed line in both figures. The modulation of the reflectivity profile of the partially sintered bicrystal at 295K is supposed to be due to voids in the interfacial region, which provide the necessary large electron density contrast for reflecting X-rays back out of the bicrystal and thus producing this interference effect. The reflectivity profile is also characterized by diffuse scattering that is a maximum at the sharp specular peak. The modulation of the diffuse scattering disappears, and the specular reflection becomes somewhat more intense after the temperature of the bicrystal is raised to 428K. The changes of the diffuse scattering profile may be related to the sintering of interfacial voids into grain boundary material, that occurs at this temperature. The diffuse scattering profile of the same

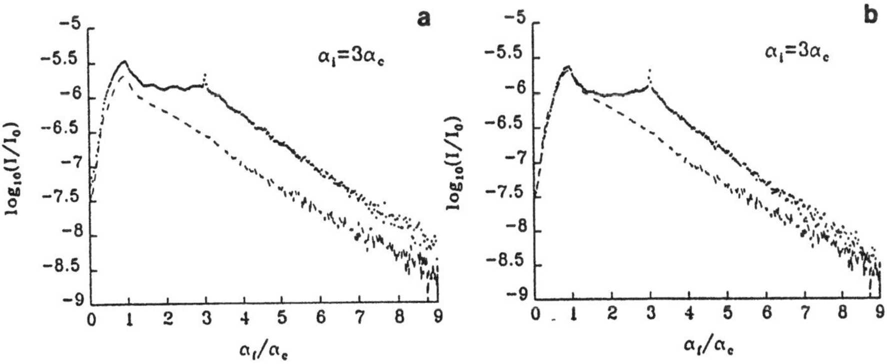

Fig. 3: The diffuse scattering for $\alpha_i = 2.82\alpha_c$ of a bicrystal with $\Theta_t = 28.1°$ at (a) 295K and (b) 428K. The profile after high temperature annealing at 634K is also shown (dashed line).

bicrystal - after its grain boundary has been removed by high temperature annealing - has a form very similar to the profile of a single thick Au single-crystal film (Fig. 1), as expected. A detailed quantitative analysis of non-specular reflectivity profiles of bicrystals with large area grain boundaries is under way.

The significant change of the diffuse scattering profiles of the same bicrystal with and without grain boundary is direct evidence that diffuse scattering from a (001) twist grain boundary has been observed.

[1] M.R. Fitzsimmons and S.L. Sass, *Acta. metall.* **36**, 3103 (1988).
[2] M.R. Fitzsimmons, E.Burkel and S.L.Sass, *Phys. Rev. Lett.* **61**, 2237 (1988).
[3] M.R. Fitzsimmons and S.L. Sass, *Acta. metall.* **37**, 1009 (1989).
[4] M.R. Fitzsimmons and E. Burkel, in progress.
[5] M.R. Fitzsimmons, E.Burkel and J.Peisl, MRS Conference Boston (1990).
[6] S. Dietrich and H. Wagner, *Z. Phys. B - Condensed Matter* **56**, 207 (1984).
[7] S.K. Sinha, E.B. Sirota, S. Caroff and H.B. Stanley, *Phys. Rev. B* **38**, 2297 (1988).
[8] G. Wallner, E. Burkel, H. Metzger and J. Peisl, *Phys. Stat. Sol. (a)* **108**, 129 (1988).

Depth Resolved Diffuse Scattering from Buried $CoSi_2$ Layers in Silicon

D. Bahr[1], *B. Burandt*[1], *M. Tolan*[1], *W. Press*[1], *R. Jebasinski*[2], *and S. Mantl*[2]

[1]Universität Kiel, Institut für Experimentelle Physik, Olshausenstr. 40, W-2300 Kiel 1, Fed. Rep. of Germany
[2]Forschungszentrum Jülich GmbH ISI, Postfach 1913, W-5170 Jülich, Fed. Rep. of Germany

Abstract. Buried single crystalline layers of $CoSi_2$ in Si with (001) surfaces were produced by ion implantation. The $CoSi_2$ is of pure A–type which until now cannot be produced by MBE for this surface orientation. X–ray experiments under grazing incidence conditions show that the transitions between the layers are rather extended and that only the inner parts of the layers are perfectly crystalline. The existence of large lattice distortions can be concluded from the diffuse scattering around the in–plane (220) reflection.

1. Introduction

For microelectronics it is very important to get contacts with low resistance and high temperature stability. Both demands are fulfilled by $CoSi_2$. In addition its CaF_2 structure is very similar to that of silicon with a lattice mismatch of only -1.2%. This allows the fabrication of 3D heterostructures.

For our experiment a silicon wafer of floating zone type with a (001)–surface was implanted with $^{59}Co^+$ ions. The energy of the ions was chosen to 200 keV and the dose to $2 \cdot 10^{17} cm^{-2}$. The implantation [1] was performed in a medium current ion accelerator (EATON NV–3204). Subsequent annealing at 750°C for 30s and at 1150°C for 10s leads to the formation of a buried single crystalline $CoSi_2$ layer of type A. The electrical resistivity amounts to 14.5$\mu\Omega$cm and the minimum yield of a channeling experiment gave 7%. RBS measurements on the sample suggest a width of 700Å of the silicide layer buried underneath a silicon layer of 900Å. Afterwards x–ray measurements were carried out at Kiel (rotating anode $\lambda = 1.54$Å), HASYLAB (beamline W1 $\lambda = 1.4$Å), and at Brookhaven (NSLS beamline X22–B $\lambda = 1.65$Å). The best results were gained at NSLS because the energy is slightly below the K–absorption edge of cobalt.

2. Reflectivity and Grazing Incidence Measurements

Reflectivity measurements were performed to determine the depth profile of the electron density. In order to get the correct values for the interface roughnesses, the diffuse scattering underneath the specular ridge was measured separately and subtracted [2]. Fig.1 shows the NSLS–measurement together with the fitted model [3] and the resulting electron density profile.

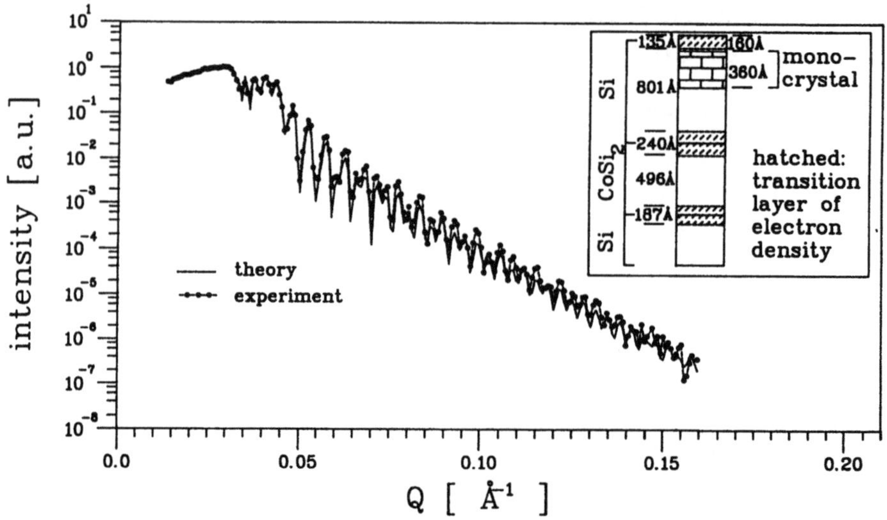

Fig.1 Reflectivity experiment at NSLS. Insertion: Resulting density profile.

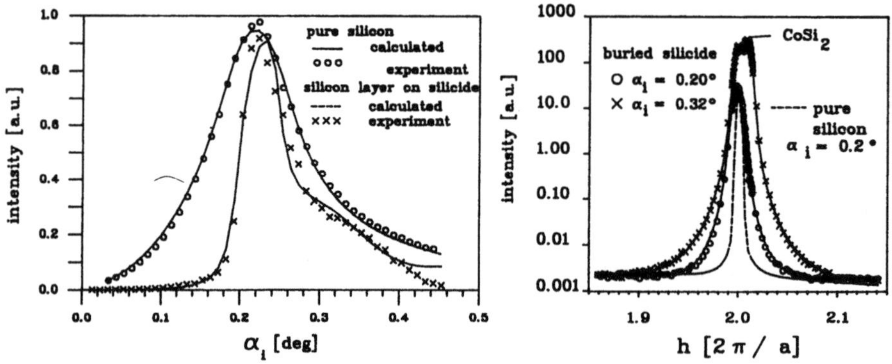

Fig.2 (220) reflection. Left: α_i dependence. Right: Diffuse scattering.

The α_i dependence of the silicon (220) and the $CoSi_2$ (220) reflection was measured at HASYLAB in an α_f resolved mode, whereas in the other cases an α_f integration was used. Due to the layered structure of the sample the known shape of a silicon (220) reflection gets strongly disturbed (fig.2, left). The result can be explained by the following model (extension of the model described in [4]):

$$I \sim |T_i|^2 \, |S_{220}|^2 \, |T_f|^2,$$

with
$$S_{220} = \frac{\exp(-iQ'_z n_s a) - \exp(-iQ'_z n_e a)}{1 - \exp(-iQ'_z a)}, \quad Q'_z = k_i^{z\prime} - k_f^{z\prime}.$$

Here the prime denotes quantities inside the crystal and a is the lattice constant.

As a result only a layer of about 360Å thickness 160Å underneath the sample surface with perfect crystalline quality is suggested. (see fig.2, left and the insertion of fig.1). n_s and n_e are the start and end numbers of this single crystalline layers.

3. Diffuse Scattering

The diffuse scattering was examined in the neighbourhood of the in-plane (220) reflections [5]. In fig. 2, right data from a reference sample (pure silicon) and of the buried silicide are shown. One can see that the scattering of the silicide can be well suppressed and made larger than that of the silicon by varying α_i. The lines represent the following expression fitted to the data:

$$I = I_{G,Si} + I_{G,CoSi} + I_{L,Si} + I_{L,CoSi}$$

with
$$I_L = \frac{I_{0L}}{\left|\frac{2(q - G_{220})}{\sigma_{FWHM}}\right|^r + 1}, \quad I_G = I_{0G} \exp\left\{-\frac{2}{\ln 2}\left(\frac{q - G_{220}}{\sigma_{FWHM}}\right)^2\right\}.$$

The term I_L is the diffuse scattering which follows a power law away from the reflection. For Huang scattering under grazing incidence conditions this results in an exponent $r = 1 \ldots 2$ [5]. This is the case for a neutron irradiated sample which has been investigated as additional reference. The thermal diffuse scattering in pure Si is weak and can be neglected for the other samples. Instead of this the silicide sample shows a power law with $r \approx 4$ and $I_L \approx I/2$. This means that the scattering must be explained in the Stokes Wilson approximation which predicts this exponent for bulk scattering [6]. From this it follows that the lattice strains must be very large and originate from extended defects.

This project was supported by the BMFT under contract number 05401ABI2.

References

[1] K. Kohlhof, S. Mantl, B. Stritzker Appl. Surf. Science **38** 207 (1989)
[2] S.K. Sinha, E.B. Sirota, S. Garoff Phys. Rev. B **38** 2297 (1988)
[3] L.G. Parrat Phys. Rev. **95** 359 (1954)
[4] H. Dosch Phys. Rev. B **35** 2137 (1987)
[5] S. Grotehans, G. Wallner, E. Burkel, H. Metzger, J. Peisl, H. Wagner Phys. Rev. B **39**, 8450 (1989)
[6] P.H. Dederichs, Journal of Physics **F3** 471 (1973)

Glancing Angle X-Ray Techniques for the Analysis of Ion Beam Modified Surfaces

T.A. Crabb[1,2] *and P.N. Gibson*[1]

[1]Institute for Advanced Materials, JRC Ispra, I-21020 Ispra (VA), Italy
[2]Department of Pure and Applied Chemistry, University of Strathclyde, Glasgow, UK

Abstract. This paper demonstrates the application of glancing angle X-ray diffraction, using a standard laboratory X-ray source, to the characterisation of ion beam modified and corroded Ni-20%wtCr steels. Also included is an example showing the potential of grazing incidence X-ray reflectivity in the study of the oxidation behaviour of corrosion resistant layers, in this case chromium.

1. Introduction

Glancing angle X-ray techniques offer a non-destructive method of surface characterisation [1]. Generally, this type of work is performed using synchrotron radiation or rotating anode laboratory sources. However, the multipurpose spectrometer [2] used to measure the spectra presented in this paper employs a conventional sealed 2kW X-ray tube. Using grazing incidence asymmetric Bragg (GIAB) geometry, it was possible to characterise as-implanted polycrystalline materials and to depth-profile the corrosion products formed after the oxidation of implanted steels.

Protection of alloys and coatings against corrosion at high temperatures is provided by protective oxide coatings, usually chromia or alumina. It is well known that the addition of reactive elements such as Y and Ce promotes the formation of a slow-growing, adherent chromia layer on Cr containing alloys, that is particularly resistant to further attack; the corrosion mechanism is still not fully understood. Ni-20wt%Cr has been implanted with Y and oxidised at 700°C. Both the as-implanted and corroded specimens have been characterised using X-ray diffraction. In order to study the effect of the implantation damage, some samples were implanted with Kr and given the same oxidation treatments.

In addition to phase identification, it is also of interest to study the formation and subsequent development of a surface oxide as function of time or temperature. In the case of deposited metallic layers X-ray reflectivity provides information on densities, layer thicknesses and interface and surface roughnesses, so that initial corrosion kinetics can be studied.

2. Experimental Arrangement and Sample Preparation

When used for diffraction, the multipurpose spectrometer previously described [2] has two modes of operation, allowing both in-plane GIXS (grazing incidence X-ray scattering), or GIAB (grazing incidence asymmetric Bragg) geometries. GIAB was used in this investigation because the materials are polycrystalline with the emphasis on phase identification as a function of depth. On our instrument the sample remains horizontal at all times, with the X-ray tube and input slit mounted on a goniometer to provide the required incident angle. A solar slit (0.2° angular divergence) and a germanium solid state detector, windowed to the CuKα radiation, are mounted on the detector arm. For reflectivity scans the solar slit is replaced by a knife edge to remove the direct beam and the detector is windowed to both the CuKα and CuKβ lines allowing simultaneous measurements at two different wavelengths.

The samples were mechanically polished to a mirror finish using 1μm diamond paste. Subsequent Kr and Y implantations were at 200KeV with a fluence of $1.0x10^{17}$ions/cm^2. Oxidation was for 8 minutes at 700°C under flowing 99%Ar/1%O_2(3lt/min).

For the reflectivity measurements Cr was sputter deposited on to optically flat substrates to thicknesses of 1200Å and 0.6μm. Ion beam modification of the 0.6μm sample was with Y at 200KeV with a fluence of $5.0x10^{16}$ions/cm^2. The 1200Å sample was oxidised for 4 minutes in air at 500°C.

3. Results and Discussion

Figures 1(a) and (b) show the diffraction spectra recorded with an incident angle of 0.5° for the as-implanted steels. In the case of Y, the surface of the sample seems to have become amorphised as indicated by the broad diffuse scattering centred near 43°; increasing the incident angle was found to increase the contribution from the sharp fcc steel peaks labelled S. The Kr implantations, on the other hand, had the effect of introducing a wide peak at 29.5°, indicating a nanocrystalline material. One explanation could be the presence of solid fcc Kr bubbles. Observations of solid Ar on implantation into Ni have already been reported [3], but have since become a matter of controversy [4]. Software developed by Gissler [5] incorporates the Wierl [6] formula, valid for randomly oriented crystallites, to determine the diffracted intensity for a specified number of atoms of a particular structure. It was found that an fcc structure with a_0=5.25Å gave the best fit to the experimental data. This agrees well with values previously reported for solid krypton inclusions [7]. Attempts to fit the data with various oxide structures were unsuccessful.

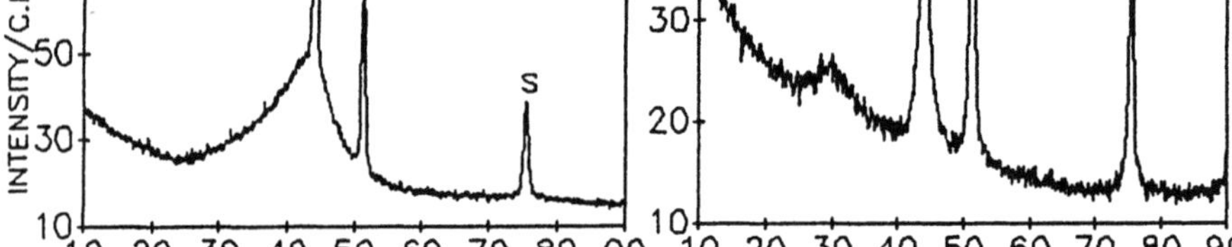

Figure 1. Diffraction from ion beam modified Ni-20%wtCr steel samples (a) Y implantation, S refers to fcc steel peaks (b) Kr implantation.

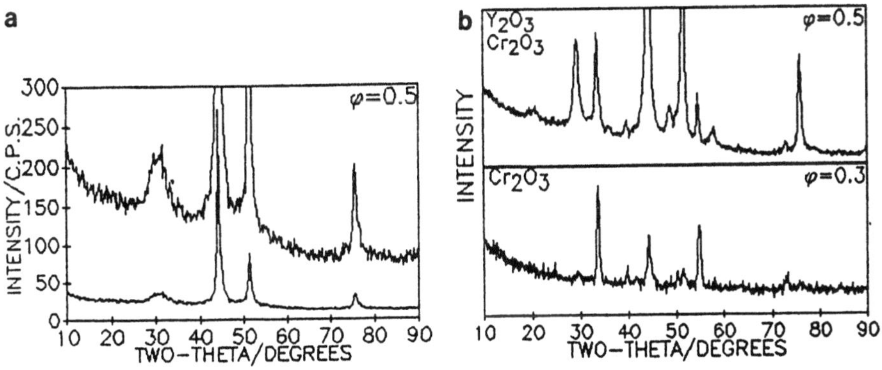

Figure 2. Oxidation at 700°C of Y implanted Ni-20%wtCr (a) 2 minutes exposure, incident angle 0.5° (b) depth profiling of sample exposed for 8 minutes, incident angles of 0.5° and 0.3°.

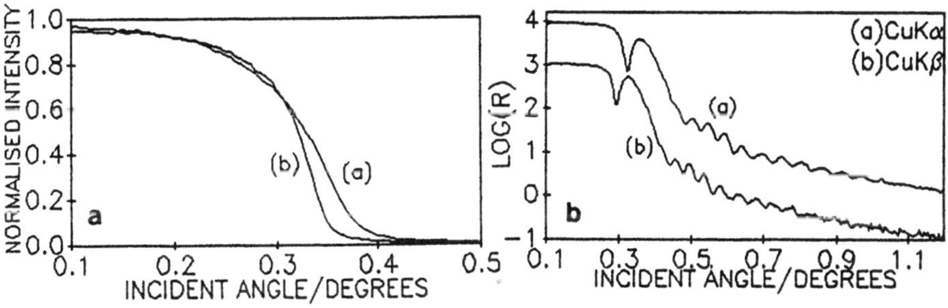

Figure 3. Reflectivity profiles from deposited Cr samples (a) 0.6μm film, as-deposited labelled "a" and Y implanted labelled "b" (b) 1200Å Cr after oxidation at 500°C for 4 minutes in air.

Figures 2(a) and (b) show the Y implanted steel after 2 and 8 minutes oxidation. After 2 minutes the amorphous phase has been replaced by a wide peak at 31°. Reducing the incident angle actually removes this contribution. We attribute this peak to the formation of nanocrystallites of hexagonal Y, segregation occurring during the initial heating of the sample. Two phases have been identified on the sample oxidised for 8 minutes with depth profiling revealing the presence of Cr_2O_3 on a sublayer of Y_2O_3, see figure 2(b).

Under oxidation the Kr implanted samples behave in a similar manner to unimplanted samples, rapidly forming a surface layer of NiO after 2 minutes. Following 8 minutes oxidation, NiO is on the surface, with Cr_2O_3 below. It was also possible to identify peaks associated with the spinel $NiCr_2O_4$, which is often present between the two oxides.

Reflectivity scans are presented in figures 3(a) and (b). Figure 3(a) is plotted on a linear scale and shows the changes which appear near the critical angle after ion beam modification of a 0.6μm Cr thin film. Figure 3(b) shows the reflectivity profile from a 1200Å oxidised sample for both the CuKα and CuKβ wavelengths; in this case the data is plotted on a log scale to emphasise the fringe region. The large peak at 0.35° is due to the formation of an oxide layer (two critical angles are apparent), the higher frequency fringes are from inteference between the surface and the Cr/glass interface. Software has been developed to fit theoretical curves to experimental data at one wavelength and is currently being modified to fit datasets at different wavelengths simultaneously. Zero point errors and uncertainties concerning the fitting can thus be removed. Accurate physical parameters such as layer thickness and density will provide information on the kinetics of the reaction and complement structural information obtained from diffraction.

4. References

[1] W.C.Marra, P.Eisenberger and A.Y.Cho, J. Appl. Phys. **50** (11), 6972 (1979)

[2] R.C.Buschert, P.N.Gibson, W.Gissler, J.Haupt and T.A.Crabb, Colloque de Physique C7 (10), 169-173 (1989)

[3] R.Khanna, A.K.Tyagi, R.V.Nandedkar and G.V.N.Rao, Scripta Metallurgica **2**, 181-184 (1986)

[4] I.F.Ferguson, Scripta Metallurgica et Materialia,**24**, 1399 (1990)

[5] W.Gissler, J.Haupt, A.Hoffman, P.N.Gibson and D.G.Rickerby, Thin Solid Films **199**, 113-122 (1991)

[6] G.M.Barrow, Physikalische Chemie, Ed. Bohmann/Vieweg & Sohn, Braunschweig

[7] L.Grabaek, J.Bohr, E.Johnson, H.H.Anderson, A.Johansen and L.Sarholt-Kristensen, Mat. Sci. & Eng. **A115**, 97-101 (1989)

Surface Analysis of Borkron Glass for Neutron Applications

M. Maaza[1;2], *C. Sella*[3], *B. Farnoux*[1], *F. Samuel*[2], *and P. Trocellier*[4]

[1]Laboratoire Léon Brillouin, (CEA-CNRS), CEN-Saclay, F-91191 Gif-sur-Yvette, France
[2]Compagnie Industrièlle des Lasers (C.I.LAS), route de Nozay, B.P. 27, F-91460 Marcoussis, France
[3]Laboratoire de Physique des Matériaux, C.N.R.S. de Meudon-Bellevue, 1 Aristide Briand, Meudon Cedex, France
[4]Laboratoire Pierre Süe, (CEA-CNRS), CEN-Saclay, F-91191 Gif-sur-Yvette, France

Abstract. Grazing angle neutron reflectometry, optical and mechanical roughness profilometry techniques have been used to study the effects of the polishing operations on the surface of Borkron Schott glass (special borosilicate glass for neutron optics applications) as the polishing tool pressure P and the mean grain size of the polishing powder Φ.

The neutron reflectivity investigations have shown that there is formation of a layer at the surface glass substrate. This layer is less dense than the bulk substrate and its thickness is around 60Å.

The optical and mechanical profilometry measurements have shown that both roughness and waviness decrease with P and Φ.

All the experimental results show a good correlation between the neutron refractive index, the thickness and the roughness of the surface layer and the waviness of the glass surface with the two mechanical polishing parameters.

The previous techniques have been completed by secondary ion mass spectroscopy and atomic force microscopy measurements.

1. Introduction

Surface microroughness is becoming more important in optical polishing for high power laser, X-ray and neutron optics. Moreover, it is recognized that the roughness of the glass substrate contributes significantly to the degradation of multilayers reflectivity. In order to improve the performance of the neutron supermirror (high multilayer neutron reflectors) [1], there has been increasing interest in the characterization and the reduction of the substrate roughness and the chemical alteration of the surface substrate.

It is also known that the physical and chemical changes at the surface of glass sample caused by polishing operations induce automatically the formation of a surface layer [2-3]. This surface layer is characterized by refractive index and a thickness and also roughness.

The purpose of this study was to gain insight into effects resulting at Borkron glass (Schott borosilicate glass) surface from polishing operations. Especially, two aspects were studied: the evolution of the characteristics of the superficial layer firstly in relation to the polishing tool pressure P which was varied from 0.5 bar to

1.5 bar and secondly to the mean size of polishing powder Φ which was varied from 0.30 μm to 1.04 μm.

For the geometrical defects, optical roughness profilometry (O.R.P.), mechanical roughness profilometry (M.P.R.) and atomic force microscopy (A.F.M.) had been used while secondary ion mass spectrometry (S.I.M.S.) has been only used to obtain the chemical profile of the glass surface. The grazing angle neutron reflectometry (G.A.N.R.) informs about the two effects. One can note that the major used techniques are non destructive.

These tests have been intended to evaluate the possible connections existing between the surface roughness and waviness, and also hydration capabilities with both polishing tool pressure P and the polishing powder granulometry Φ on the surface of Borkron glasses.

2. General experimental procedure

2.1 Surface technique investigation

In the GANR technique, the wave nature of the neutron particle is considered. Neglecting the absorption effect, the optical neutron-matter interaction is described by a refractive index which is related to the composition of the glass and defined as [4] :

$$n_{ne} \approx 1 - \lambda^2 (\Sigma \mathcal{N}_i b_i)/2\pi , \qquad (1)$$

where λ is de Broglie associated neutron wavelength, $\mathcal{N}_i$ is the number of scatterer atoms per unit volume of species i, b_i which characterizes the interaction is the coherent nuclear scattering length. We measure the term $\mathcal{N}b = (\sum_i \mathcal{N}_i b_i)$ which is called the nuclear coherent scattering length density.

Currently, in this technique, the time of flight (T.O.F.) is applied. The neutron reflectometer DESIR (Léon Brillouin Laboratory, C.E.N.Saclay, France) was used. It was described in [5]. The glancing angle was kept constant ($\theta_0 \approx 1.5°$) and the neutron wavelength is varied from 3Å to 30Å. The specular neutron reflectivity will be plotted versus the normal neutron wave vector component $Q = 2\pi \sin\theta_0/\lambda$.

For simulating the experimental neutron reflectivity curves, one has to consider an homogeneous thin layer at the top of the bulk glass. From neutron optical point of view, this surface layer is characterized by a refractive index n_{sl} (or neutron scattering length density $\mathcal{N}b_{sl}$) and a thickness d_{sl} which is deposited on the bulk substrate whose refractive index is n_s (or a neutron scattering length density $\mathcal{N}b_s$). The neutron reflectivity profile for this model structure may be calculated exactly using the optical matrix method for the reflection of light polarized perpendicular to the reflection plane [6]. In conclusion, two parameters were then deduced from the GARN measurements : the neutron scattering length density Nb_{sl} of the surface layer and its thickness d_{sl}.

In the ORP measurements, the surface profile is point by point displayed of the relative height of the surface being examined. The vertical resolution was of order 1Å and a lateral resolution is 0.79 μm. The tested profile length is 0.333 μm. Two geometrical parameters were deduced from these measurements : the rms roughness σ_{rms} and the peak-to-valley roughness σ_{P-V}.They represent the average root-mean-square deviation of the profile and the profile irregularity height respectively.

In the MRP measurements, the scanned length was 200μm, the scan speed was 2μm/s. The vertical resolution is around 1 Å, thought the lateral resolution is of the order of 1μm. The parameters deduced from the Tencor P-1 measurements are the rms and peak-to-valley waviness of the surface w_{rms} and w_{P-V} and also rms and peak-to-valley roughnesses respectively. From the integrated signal, the roughness and the waveness are separated using selectable cutoff filters.

Atomic force microscopy measurements were realized on a Nanoscope atomic force microscope II on the glass samples. This last technique permits to perform nanometer profiles. The aFM imagings were done at room temperature and atmospheric pressure.

2.2 Sample Preparation

The mechanically polished types of Borkron glass (200x90x15 mm^3) have been been tested are commonly used as substrates for neutron mirrors [7-9]. These types of glass are of particular interest because they contain boron atoms (≈13% oxide in weight) having a strong absorption cross section . They permit to reduce the background around the neutron guides and the spectrometers. They also play a role in biological shielding. Their surface state depends strongly on the polishing method.

The Borkron substrates were prepared using a similar procedure. After grinding with free abrasive (Silicium carbide, the mean grain diameter was $\Phi \approx 14$ μm), the samples were first polished on a felt polisher during 2 hours with cerium oxide suspension. The initial pH value was around 6.5. After this first polishing operation, Borkron samples were cleaned by R.B.S. solution at 50°C, H_2O at 25°C

Table 1 : Wyko profilometry, Tencor profilometry and neutron reflectivity experimental parameters of different types of Borkron glass

Samples	BK12	BK14	BK16	BK38	BK18	BK12
Pressure [bar]	0.5	1.0	1.5	0.5	0.5	0.5
Grain size [μm]	0.75	0.75	0.75	0.30	0.75	1.04
Thickness of surface layer [Å]	50	50	70	40	63	85
Neutron density $\mathcal{N}b$ of surface layer [10-6Å^{-2}]	1.1	0.7	0.5	2.2	2.8	3.0
Optical peak-to-valley roughness [Å]	25.2	26.7	33.3	31.4	29.1	45.9
Optical rms roughness [Å]	4.09	5.09	6.72	4.78	5.17	8.84
Mechanical peak-to-valley roughness [Å]	52	79	45	47	48	87
Mechanical rms roughness [Å]	5.9	5.9	5.9	5.9	5.9	6.9
Mechanical peak-to-valley waviness [Å]	59	70	212	49	90	102
Mechanical rms waviness [Å]	12.9	14.9	46	10.9	16.9	23.8

and ethylic alcohol respectively. After this first approach, the samples were superpolished with a polytron pad covering about 33% of a 30 shares perbunan sheet glued onto a flat tool. This polishing tool was continuously feed by a deionized water (pH ≈ 6.5) containing 30% (in volume) of cerium oxide in aqueous suspension. Its temperature was held around 30°C during the whole polishing operation. After the superpolishing, the samples were rinsed by ethylic alcohol and conserved in a special paper for avoiding a new atmospheric alteration.

The types of mechanical polished Borkron glass were processed in two series. At first, the polishing pressure was varied and the grain size of the polishing powder is kept constant ($\Phi \approx 0.75$ µm) for 3 samples (Table 1). In the second case, we have proceeded inversely the polishing pressure ($P \approx 0.5$ bars) was kept constant and the granulometry was varied for the 3 latter samples.

3. Experimental Results

GARN, ORP and MRP experimental results are summarized in Table 1. According to this table, it is suggested that there is a correlation between the characteristics of the glass surface with the polishing parameters P and Φ.

3.1 GANR Data

Typical experimental GANR curves of polishedsamples are shown in Figs 1 and 2 which represent a decimal logarithmic plot versus normal neutron wave vector component $Q=2\pi\sin\theta_0/\lambda$. We discern two regions : the total reflection plateau ($Q<Q_c=2\pi\sin\theta_0/\lambda_c$) and the Fresnel region $Q \geq Q_c$. In this last region, the experimental reflectivity is always less than the theoretical one which corresponds to the bulk substrate. We conclude the existence of an antireflection surface layer whose characteristics ($\mathcal{N}b_{sl}$ and d_{sl}) were deduced by simulation and are given in Table 1. One can note that the Debye-Waller model is not sensitive in this Q range.

According to this table, the surface layer neutron scattering length density $\mathcal{N}b_{sl}$ is always less than the bulk one in all the cases (the theoretical value of the neutron scattering length density of bulk substrate was extrapolated from its chemical composition ($\mathcal{N}b_{bulk} \approx 4.2\ 10^{-6}$ Å^{-2}). It appears that $\mathcal{N}b_{sl}$ increases with the inverse of the polishing pressure at constant granulometry ($\Phi \approx 0.75$µm) and with the granulometry Φ at constant polishing tool pressure ($P \approx 0.5$ bar) . The reason for this will be discussed later in connection with the chemical surface changes. Accurately, the variation of the surface layer scattering length density $\mathcal{N}b_{sl}$ is less important at variable granulometry. The ratio ($\mathcal{N}b_{sl}/\mathcal{N}b_{bulk}$) varies between 12%-26% at varied pressure and 52%-72% at varied granulometry.

The surface layer thickness d_{sl} does not vary significantly with the polishing pressure P. It average value is around 57Å. In contrast, it varies proportionally to the granulometry Φ at fixed pressure ($P \approx 0.5$ bar).

3.2 ORP, MRP and AFM Data

The ORP measurements indicate that the rms and the peak-to-valley roughness increases rapidly with the pressure P as well as with the grain size Φ. The ORP roughness varies from 4.8 Å to 8.9 Å (P = constant) against 4.1 Å to 6.7 Å

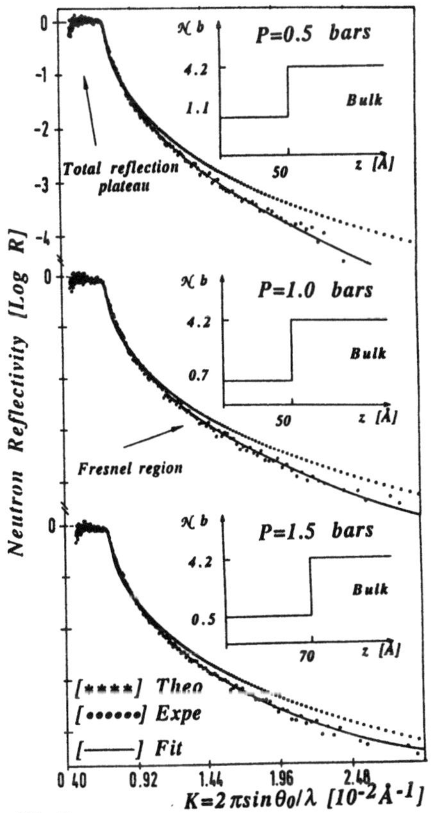

Fig 1

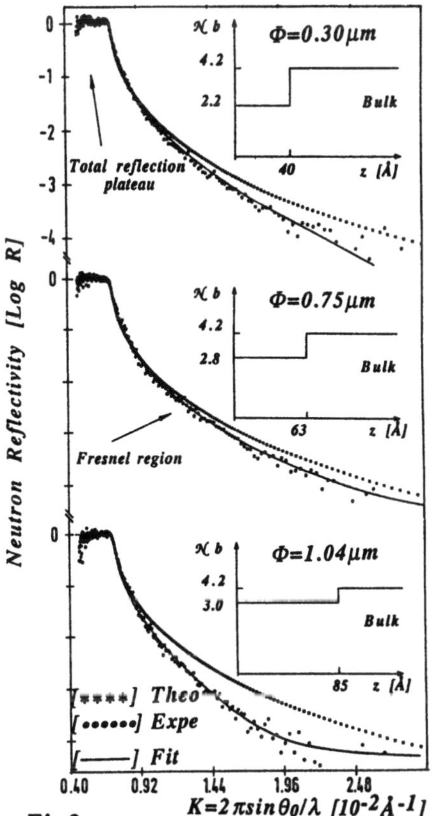

Fig 2

Fig 1. Neutron reflectivity profile of Borkron glass samples (Logarithmic scale) versus normal neutron wave vector component $Q = 2\pi\sin\theta_0/\lambda$. at constant granulometry ($\Phi \approx 0.75$ μm) and various polishing tool pressure P.The models used to simulate the reflectivity profiles are shown in the insert.

Fig 2. Neutron reflectivity profile of Borkron glass samples (Logarithmic scale) versus normal neutron wave vector component $Q = 2\pi\sin\theta_0/\lambda$.at constant polishing tool pressure ($P \approx 0.5$ bar) and various granulometry Φ.The models used to simulate the reflectivity profiles are shown in the insert.

(Φ=constant). The peak-to-valley roughness varies from 31.4 Å to 45.9 Å (P=constant) against 25.2 Å to 33.3 Å (Φ = constant).

The Tencor measurements are more sensitive to the waviness. Accurately, the table 1 shows that the variation of the waviness is more pronounced with at varied polishing pressure P at varied granulometry Φ. The peak-to-valley waviness varies from 59 Å to 212 Å at constant P and from 49 Å to 102 Å at constant Φ.

The AFM scanning micrograph showed that the surface was porous ; the average value of this microporosity was found to be around 150 Å. These tests have also shown the existence of some cracks caused during the polishing operations whose average depth is around 20 Å.

3.3. SIMS Data

A chemical profil measured by SIMS (Fig.3) was realized. A gold conductive layer was deposited on the glass surface in order to avoid the charging effects. The secondary ion mass spectrometry result shows clearly the greater depletion of sodium at the surface and the depth penetration of hydrogen. The Fig. also indicates little or no effects on the other oxides at the surface. Depth scale was calibrated by measuring the crater depth. The depleted sodium region appears to extend 400 Å into the glass. However, the region where the variation of sodium concentration is the steepest, corresponds to a thickness of 100 Å.

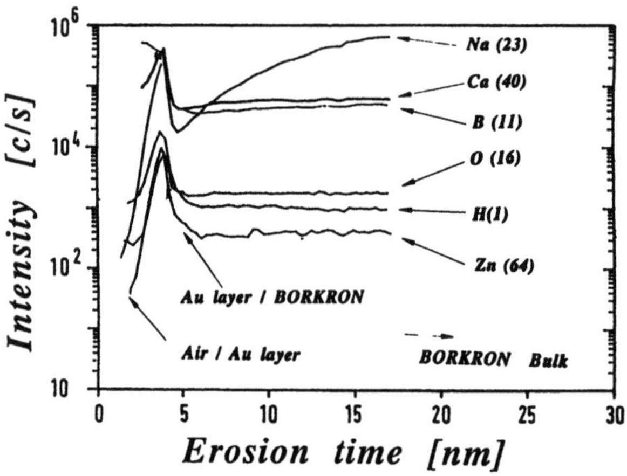

Fig 3. Secondary ion mass spectrometry depth profiles of borkron glass sample (BK 12). Note the significant depletion of Na at the surface and the improvement on H.

4. Conclusion

From the present study, the following conclusions can be drawn:

[a] The grazing angle neutron reflectometry can be used to characterize accurately the glass surface structure. It is extremely sensitive and non destructive surface investigation technique. The minimum probed thickness is around 40 Å.

[b] A less dense thin film was found at the surface of mechanically polished glass. The altered thickness at the surface have values falling within the range previously determined for these types of glass by other technique. Its average thickness is 60 Å for Borkron glass.

[c] The polished surface present a depletion on alkali ions. This effect creates a defect in electric charges and induces an ionic exchange especially between the alkali ions (Na^+, Ca^{2+}) and H^+ or/and H_3O^+.

[d] These results show that this hydration rate at the surface increases with both the polishing pressure tool and the inverse of the grain size This hydration phenomenon suggests to be more enhanced by the polishing pressure than the powder grain size at least in this range of values of P and Φ. The low neutron scattering length density of the surface layer is possibly evidence for hydronium ions H_3O^+ rather than protons ions H^+.

[e] The weak values of the surface layer neutron scattering length density suggest that the surface is porous, results were confirmed by AFM measurements.

[f] The surface roughness is mainly influenced by the granulometry of the polishing powder. The results show that the mechanical polishing of Borkron substrate at an optimal weak value of granulometry ($\Phi \approx 0.3\mu m$] and weak tool pressure ($P \approx 0.5$ bar) allow to obtain a best surface quality (weak altered thickness of 50 Å and a rms roughness of 5 Å and a waviness of 59 Å).
[g] There is an equivalence between the peak-to-valley measured by both ORP and MRP and the thickness d_{s1} of the surface layer measured by GANR. It was found that the surface layer thickness is approximately 2 times higher than the peak-to-valley measured by ORP and in order of the peak-to-valley roughness determined by MRP. This is due to the different resolution of the two techniques.

The results show that the structure (altered thickness, hydration effect, roughness and waviness) of the glass surface particularly for Borkron glass is correlated to the mechanical polishing parameters : the polishing tool pressure and the polishing powder grain size. Finally, the optimal polishing parameters deduced from the present work are P=0.5 bar and Φ=0.30μm.

Acknowledgements

This work was supported by the Commissariat à l'Energie Atomique (C.E.A.) and the Compagnie Industrielle des Lasers (C.I.L.A.S). The authors would like to thank Drs W. Kaiser, M. Miloche and G. Coddens for their help.

References

[1].Schaerft O, Physica B, 156 and 157, pp 631-638, North-Holland, Amsterdam, 1989.
[2]. Nevot L., Croce P., Rev. Phys. Appl;, 15, 761-779, 1980.
[3]. Yokota H. , Sakata H. , Nishibori M. , Kinosita K. ,Surface Science,16, 265-274, 1969.
[4] Fermi E., Zinn W.H. Phys. Rev. 70, 103, 1946.
[5]. Farnoux. B., Mat.Res.Soc. Symposium Proceedings, Vol. 166., 95-101,1990.
[6]. Sears. V.F., Neutron Optics., Clarendon Press. Oxford., 1988.
[7] Grebenscikov I., Sot. Reconstr. Nauka, 2, 22, 1935.
[8]. Kaller A., Silikaty,7. 380-390, 1956.
[9]. Bourgeaux M., Peyches I., Colloque sur la nature des surfaces vitreuses polies, Paris , published by U.S.C.V. Charleroi-Belgique, 1959.

X-Ray Bragg Reflectivity of ErAs Epitaxial Films

P.F. Miceli, C.J. Palmstrøm, and K.W. Moyers

Bellcore, 331 Newman Springs Rd., Red Bank, NJ 07701, USA

Abstract. We summarize recent x-ray scattering work on ErAs buried epitaxial layers grown on [001]GaAs. Both the low angle and high angle reflectivity is measured and modeled for films ranging in thickness from 2 atomic layers to 1000Å. This extended range of reflectivity provides useful information beyond that which can be obtained from only low angle reflectivity. We find that the character of the film thickness fluctuations changes with increasing misfit dislocation density. The transverse scattering across the Bragg reflectivity shows a two component line shape which is related to rotational fluctuations associated with misfit dislocations.

1. Introduction

There has been extensive use of x-ray and neutron reflectivity performed at low angles to investigate laminar materials [1]. Because these measurements are at low angles, length scales longer than interatomic distances are probed and the results are independent of crystal structure. For a crystalline layer, the low angle reflectivity will be replicated also at higher angles about the Bragg positions. This *Bragg reflectivity* can provide useful information since the relative contributions to the total scattering coming from other layers in a laminar structure will depend on the particular order of Bragg reflection. One can exploit the behavior of amorphous layers, forbidden reflections and films which are incommensurate with the substrate (using non-specular Bragg reflectivity). Thus, measurements for several orders of Bragg reflection as well as at low angles should be able to provide highly quantitative information regarding buried interfaces and would be particularly useful in the study of epitaxial growth.

The ErAs/GaAs system is ideally suited for reflectivity studies of epitaxial growth since the full range of film thicknesses where lattice relaxation [2] takes place is accessible (critical thickness ~70Å). Furthermore, Er has a large x-ray scattering cross section and there is no interdiffusion at the interfaces because of the high heat of formation and melting temperature of ErAs.

The samples were prepared by molecular beam epitaxy, as described elsewhere [3]. ErAs has the rock salt structure and grows along [001] on the same orientation of GaAs. The samples were capped by an overlayer to prevent reaction with the ambient atmosphere. X-ray scattering was performed using $MoK_{\alpha 1}$ radiation from an 18kW rotating anode generator. A Ge(111) monochromator provided an incident beam having an angular divergence of ~0.002° and ~1° in and out of the diffraction plane, respectively. The scattered radiation was measured using either a Ge(111) analyser crystal or two slits having a 0.05° acceptance in front of the detector.

2. Scattering Model

A general model for the specular reflectivity is obtained by summing the contributions to the scattering from each layer in the system,

$$I_{Spec} \propto |<A>|^2 = |\sum_{l=0}^{layers} <A_l e^{i\phi_l}>|^2,$$

where A_l is the scattering amplitude of the lth layer, ϕ_l is the phase shift to that layer, and <...> indicates an average over the in-plane dimensions of the film. For breveity, we present only the main results and a more detailed discussion will appear elsewhere [4]. If the layer thickness fluctuations of the various layers occur independently, then the problem reduces simply to evaluating the average of phase factors,

$$V_l = <e^{iQ_z d_l N_l}> = \sum_{n=0} P_n e^{iQ_z d_l n},$$

where Q_z is the momentum transfer normal to the film, d_l is the interplanar spacing of the lth layer, and N_l is the number of atomic planes. P_n is the probability to find n planes in the lth layer. Because the sum is over integers, the translational symmetry of buried crystalline layers is preserved and Bragg peaks will be allowed for. If these averages were done as continuous integrals, only the low angle reflectivity could be modeled. It now remains to specify P_n, as discussed below.

2.1 Monolayers

Fig. 1 shows three orders of reflectivity data obtained for a nominally 2 monolayer (ML) film of ErAs buried under 500Å of GaAs. In this case, the 0th order reflectivity does not directly measure the thin ErAs layer, but rather it is sensitive to the gap between two GaAs slabs -- a result not suprising in view of the relatively small quantity of ErAs. By contrast, the GaAs scattering is much reduced at the (002) so that the ErAs layer produces a broad scattering feature. At the (004) the GaAs scattering again dominates. Because there are only a few ErAs atomic layers, we have fitted the above model to the data to directly obtain the coefficients, P_n. We find P_1 = 0.0±0.02, P_2 = 0.04±0.08, P_3 = 0.41±0.07, P_4 = 0.19±0.07, P_5= 0.07±0.04, and P_n = 0 for n>5. This gives an average coverage of 2.4ML (±0.3) consistent with growth conditions, but P_0 = 0.3±0.2 is the fraction of *no coverage*, indicating that there are "pin-holes" in this film. It turns out that all three orders of measured reflectivity were required to obtain quantitative results, particularly for P_0. Furthermore, the solid curve showing the results of the fit demonstrates that our scattering model can reproduce both the low angle as well as the Bragg reflectivity.

The inset shows data for nominally 5 ML ErAs buried by 500Å of GaAs where there are now more intensity oscillations due to the thicker ErAs layer. A least squares fit (solid curve) to the data yields (to within ±0.03) P_3 = 0.01,

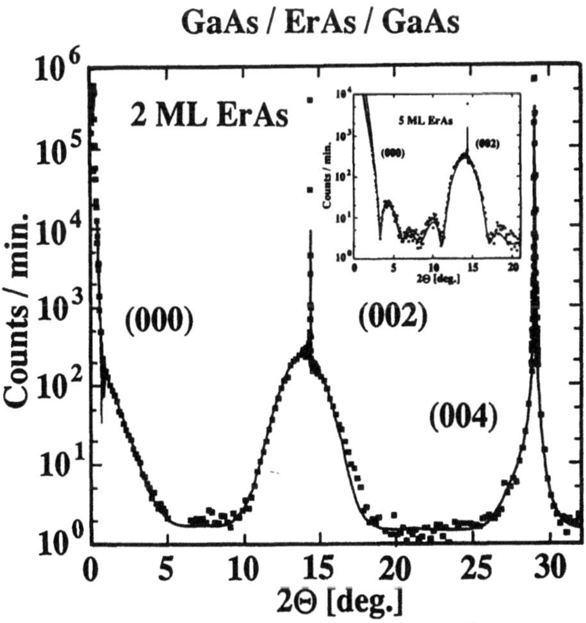

Fig. 1: 2 ML of ErAs buried beneath 500Å of GaAs. Inset shows 5 ML ErAs. Solid curves are a least squares fit of the model to the data.

P_4 = 0.41, P_5 = 0.55, P_6 = 0.03, and all other P_n = 0 are constrained. Thus, the average coverage is 4.6 ML with nearly equal amounts of 4 and 5 ML regions, indicating that the film is atomically uniform. This is consistent with the observed behavior of the RHEED oscillations during film growth which are poor at 2 ML but are largest around 4 or 5 ML.

2.2 Binomial Fluctuations

Data for 35Å of ErAs buried beneath 200Å of amorphous Si are shown in fig. 2 where intensity oscillations appear both at low and high angles due to the thin ErAs layer. The solid curve is the result of a least squares fit to the data assuming binomial layer thickness fluctuations which are a natural description of epitaxial growth since an atom will occupy a site with some probability, p, or it will not occupy it with probability, 1-p. These fluctuations will affect both the amplitude and the phase of the intensity oscillations which can be calculated analytically. We find,

$$V_l = \left[(1-\sigma_l^2/N_l)e^{iQ_z d_l} + \sigma_l^2/N_l\right]^{\frac{N_l^2}{N_l-\sigma_l^2}},$$

where here N_l is the average number of layers and σ_l^2 is the variance. In the limit of large N_l, these become Gaussian thickness fluctuations. Note that this model fits the low angle reflectivity where the amorphous layer scattering is important as well as the high angle reflectivity which is dominated by the ErAs crystalline scattering.

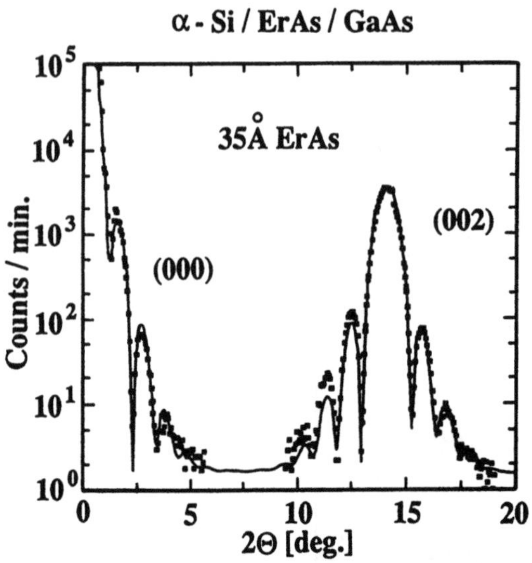

Fig. 2: 35Å ErAs buried under 200Å of amorphous Si. The model (solid curve) assumes binomial thickness fluctuations and fits both the low angle scattering where the amorphous contribution is important and the high angle data which is due to the crystalline scattering. The interface roughness of the substrate is found to be 1 ML. σ is 1 ML for the ErAs layer and 10Å for the amorphous Si cap layer.

2.3 Thick Films

The (002) Bragg reflectivity is shown in fig. 3 for a 500Å film where the lattice relaxation is essentially complete and the transverse peak width behaves like a mosaic -- ie. constant transverse angular width is observed for all (00L). The dashed curve, which is calculated by assuming the above specular reflectivity model with binomial thickness fluctuations, provides a poor description of the

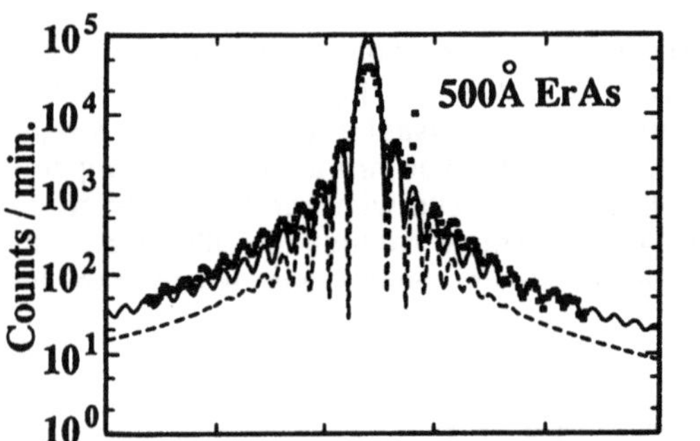

Fig. 3: 500Å of ErAs buried by amorphous Si. The dashed curve is calculated from a specular reflectivity model using binomial fluctuations with $\sigma = 10$Å. The solid curve uses the integrated reflectivity with exponential fluctuations (f=0.93).

data in two important ways: (1) the "tails" of the calculated reflectivity are too low and (2) the intensity oscillations decay much too rapidly -- a problem which cannot be resurrected by simply changing the size of the thickness fluctuations.

The increased intensity in the tails can be accounted for by assuming that we are measuring transversely integrated reflectivity rather than specular reflectivity. The distinction lies in the way the averages are performed and the integrated intensity always provides more scattering in the tails, $I_{Int} - I_{Spec} = \langle |A|^2 \rangle - |\langle A \rangle|^2 \geq 0$. This may be understood by considering the standard picture of mosaic blocks which are randomly oriented and, thus, automatically perform the Q_x and Q_y integration as we measure the reflectivity along Q_z. Although the concept of mosaic blocks is probably much oversimplified, note that a mosaic sample contains no true specular scattering as it is not resolution limited.

By assuming exponential thickness fluctuations, $P_n \propto f^n$ $(0 \leq f \leq 1)$, the rate of decay of the intensity oscillations is algebraic and this fits the data rather well as shown by the solid curve in fig. 3, which also includes the above mentioned integrated method of averaging intensities. The slight difference between the model and data near the center of the Bragg reflectivity is due to inhomogenous strain which is not yet included in our model. It is interesting that the basic character of the film thickness fluctuations change depending on whether the films are psuedomorphic or relaxed -- this might be a consequence of the change in surface free energy associated with the presence of a high density of misfit dislocations.

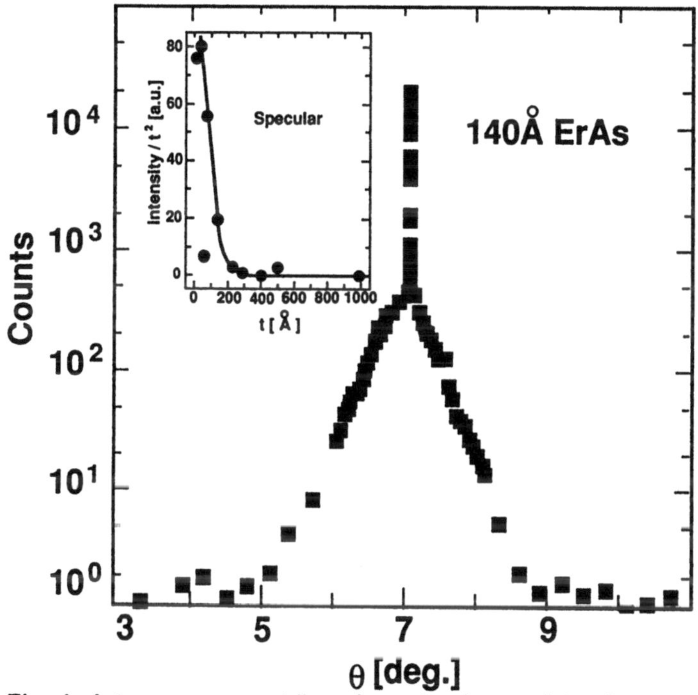

Fig. 4: A two component line shape is observed for the transverse scattering taken at the center of the (002) reflectivity of a 140Å ErAs film. Inset shows the film thickness dependence of the intensity of the sharp central component.

3. Transverse Bragg Reflectivity

A transverse scan across the center of the (002) Bragg reflectivity for a 140Å ErAs film is shown in fig. 4 where a two component line shape is observed. There is a sharp, resolution limited (specular) component which suggests some character of flatness over a correlation range of several microns and a broad diffuse component which is due to mosaic-like rotational fluctuations. With increasing film thickness, the intensity of the sharp component decreases (see inset) and the broad component increases, which evolves into a conventional mosaic as the dislocation density increases. The presence of the sharp central component suggests that the out-of-plane rotational fluctuations are bounded at infinite length scales for the thinner films. However, a similar central component is never observed for in-plane reflections, such as the (222), implying a relatively weak binding between ErAs and the GaAs substrate as will be discussed in a future publication [4].

References

1. J. Als Nielsen, in *Structure and Dynamics of Surfaces II. Phenomena, Models and Methods*, ed. by W. Schommers and P. Blanckenhagen, Topics in Current Physics, Vol. 43 (Springer Verlag, Berlin 1987). C. F. Majkrzak and G. P. Felcher, MRS Bull. **15** (11) 65 (1990).
2. P. F. Miceli, C. J. Palmstrøm and K. W. Moyers, Appl. Phys. Lett. **58**, 1602 (1991).
3. C. J. Palmstrøm, N. Tabatabaie and S. J. Allen, Appl. Phys. Lett. **53**, 2608 (1988). C. J. Palmstrøm et. al., J. Vac. Sci. Technol. **B 7**, 747 (1989).
4. P. F. Miceli and C. J. Palmstrøm, unpublished.

Measurement of Magnetic Field Penetration Depth in Niobium Polycrystalline Films by the Polarized Neutron Reflection Method

L.P. Chernenko, D.A. Korneev, A.V. Petrenko, N.I. Balalykin, and A.V. Skripnik

Laboratory of Neutron Physics, Joint Institute for Nuclear Research, Head Post Office, P.O. Box 79, SU-101000 Moscow, USSR

Abstract. The values of the superconducting penetration depth (s.p.d.) measured on "thin" (255 nm) and "thick" (700 nm) niobium films are presented. The obtained values of s.p.d. in our experiments at T=4.9 K are 145(+/-15) nm for a "thin" film, and 90(+/-10) nm for a "thick" film. It is essentially different from that one for bulk niobium (43 nm at T=4.6 K after G.P. Felcher et al., Phys.Rev.Lett. **52**, 1539 (1984)). The experiment has been carried out on the polarized neutron spectrometer at the IBR-2 reactor in Dubna.

As shown in the literature [1], the specular polarized neutron reflection may be used as a direct method of measuring the absolute value of the magnetic field penetration depth of a superconductor (s.p.d.). However, the s.p.d. as determined by measuring the direct current in Josephson junctions in a weak magnetic field [2] has given a value in niobium films (91 nm, T=4.6 K) which is well over that obtained by polarized neutron reflection [1] (43 nm, T=4.6 K). The present experiment was started to throw light on causes of this difference.

We carried out experiments on two niobium films of various thickness and roughness prepared with the same sputtering technique. The experiments have been conducted on the spectrometer of polarized neutrons (SPN) in a reflectometry mode [3] at the IBR-2 reactor. Two measurements have been carried out sequentially. At room temperature the specular reflection spectra were measured to obtain neutron-optical parameters (thickness, roughness) of films. Then we carried out the measurements once more at T=4.9 K in the field of 500 Oe. Before the measurements were performed, the reflectometer was turned to the grazing angle θ=4.0 mrad with $\Delta\theta/\theta$=0.025. For data handling we used the new method of calculation of the reflection factor described in [4]. The method consists of replacing the continuous one-dimensional neutron-optical potential of the film with the discrete series of Fermi quasi-potentials to model the reflection of plane waves from inhomogeneous media. Unlike the traditional approach [1], we took roughness into account by introducing in the calculation of the reflectivity increasing amplitudes of Fermi potentials. These were distributed according to the Gaussian error curve with dispersion equal to the roughness parameter (s) square. Actually the results of the two techniques are in excellent agreement for pure nuclear potentials. The two methods to evaluate roughness become different only when both nuclear and magnetic potentials are present [4].

In view of the geometric arrangement of the sputtering apparatus, the "thin" film of 28x50 mm^2 (sputtered on a silicon plate) had a graded thickness with an average value of 255(+/-15) nm, σ=0.5 nm. The "thick" film - 700 nm in thickness - was obtained by slanted sputtering on the ceramic substrate (95% Al_2O_3), σ=8 nm. The critical temperature for both films was 8.95 K. Since a decrease of the critical temperature of

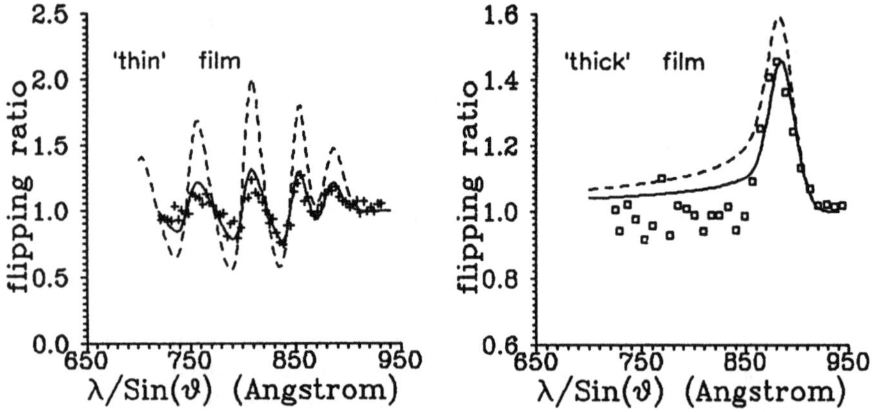

Figure. Experimental values of the flipping ratio: crosses are for the film 255 nm in thickness; squares are for the film 700 nm in thickness. Lines correspond to theoretical calculations considering the corrections for polarization and resolution of the reflectometer. Dashed lines are for the s.p.d. of 50 nm.

niobium films is expected at a film thickness lower than 200 nm the samples seemed satisfactory. This was confirmed by the analysis of the composition of initial material for niobium sputtering, and the films themselves, using the neutron activation analysis. The analysis did not reveal noticeable amounts of impurities in the films.

To analyse the experimental data we used the diamagnetic profile of a superconducting film according to London's local electrodynamics of superconductors. An extra ingredient of the model was the assumption of the existence of some "dead" layer in niobium close to a substrate, which at T=4.9 K does not pass to a superconducting state.

The figure shows the relation of reflection factors of neutrons with opposite polarization (flipping ratio) depending on a normal component of a neutron wave length. If the model does not consider the "dead" layer on the boundary with a sublayer, the experimental spectra are described by theoretical curves (see fig.) at the value of s.p.d. equal to 145(+/-15) nm for a "thin" film, and 90 (+/-10) nm for a "thick" one, respectively. The value of a s.p.d. equal to 43 nm [1] describes our data adequately only for a "thin" film in a model with a "dead" layer 100 nm in thickness.

The analysis of systematic factors leading to a possible shift of s.p.d. has shown that the obtained values at T=4.9 K of 145 nm and 90 nm are the low values of s.p.d. for the thin films investigated. We believe that the hypothesis of a "dead" layer 100 nm in thickness, bringing the value of s.p.d. to that of 43 nm (T=4.9 K) is not well based for two reasons. Firstly, the analysis of the depth composition of a film by backward scattering of accelerated ions of helium with an energy from 3.0 Mev to 3.2 Mev has shown that there is neither oxygen nor other impurities across the whole width of the thin film. Secondly, the films 100 nm in thickness at T=4.9 K are superconducting.

Conclusions

We conclude that our measurement of s.p.d. by neutron reflectometry method on the niobium polycrystalline films gave the s.p.d. that is different from the s.p.d. in bulk niobium. At T=4.9 K the s.p.d. for the "thin" film and that for the "thick" film are 145 nm and 90 nm, respectively.

The studies of the film contents by neutron activation analysis and backward scattering of He ions did not confirm hypothesis of the existence of "dead" layer in the "thin" film, which was introduced to explain the experimental data with s.p.d. of bulk niobium.

The authors wish to thank Dr. V.Nazarov for the activation analysis of samples. Thanks are due to Dr.A.Kobzev and Dr.D.Shirokov for backward scattering spectrometry.

References

[1] G.P.Felcher, R.T.Kampwirth, K.E.Gray and R.Felici, Phys.Rev.Lett., **52**, 1539 (1984).

[2] M.Russa, J.Low Temp. Phys., **50**, 301 (1983).

[3] D.A.Korneev, V.V.Pasyuk, A.V.Petrenko and E.B.Dokukin, "Neutron reflectivity studies on superconducting, magnetic and absorbing thin films on the polarized neutron spectrometer at the pulsed reactor IBR-2", (see this issue).

[4] D.A.Korneev and L.P.Chernenko, JINR Rapid Communication 4[30]-88 and Poverhnost'. Fizika, himija, mehanika. **9**, 61 (1990) (in russian)

Neutron Reflectivity Studies on Superconducting, Magnetic and Absorbing Thin Films on the Polarized Neutron Spectrometer at the Pulsed Reactor IBR-2

D.A. Korneev, V.V. Pasyuk, A.V. Petrenko, and E.B. Dokukin

Laboratory of Neutron Physics, Joint Institute for Nuclear Research, Head Post Office, P.O. Box 79, SU-10100 Moscow, USSR

Abstract. The Time-of-Flight Polarized Neutron Spectrometer (SPN-1) is installed at the IBR-2 reactor (JINR, Dubna). The possibility of using this instrument for the reflectometry of thin films is described. Some experiments performed in the reflectometry mode are reviewed.

1. Introduction

The phenomenon of the thermal neutron total reflection is well known from the forties. It was first studied by E. Fermi and W. Zinn [1] and since was applied to the production of neutron guides and polarizers and for the determination of the neutron scattering lengths. But the systematic application of the thermal neutron specular reflection for the investigation of neutron-nucleus and magnetic potential profiles of thin films, interfaces and surfaces have been made relatively recently. Over the last decade neutron reflectometers have been created at Argonne National Laboratory [2], Rutherford Appleton Laboratory [3], CEN-Sacley [4], KFA-Julich [5], also D17 at ILL-Grenoble [6] has been used extensively to do reflection experiments.

The Spectrometer of Polarized Neutrons (SPN-1) is installed at the pulsed reactor IBR-2 at the Joint Institute for Nuclear Research. It was designed in 1985 for the investigation of the depolarization processes of the neutrons transmitted through magnetic inhomogeneous systems such as ferromagnetics and superconductors. After the reconstruction in 1987 one has got the possibility of using this spectrometer in the reflectometry mode. The technique was applied to the study of surface and interfacial problems in the following fields: the processes of the magnetization reversal of FeCo thin films, the investigation of the titanium film aging, the measurements of magnetic field penetration depth in high Tc and classical superconducting films.

In this paper the parameters of the instrument in the reflectometry mode are described and some results from the fields of surface magnetism and neutron absorbing thin films are presented to illustrate the reflectometer capabilities.

2. Instrument Description

A schematic drawing of the SPN-1 is shown in Figure 1. A double disk chopper suppresses the background from the satellite reactor pulses and also defines the neutron wavelength band from 0.07 to 1.2 nm. The chopper discs rotate in opposite directions with velocities which can be varied within the range from 5 to 15 Hz.

The conical neutron guide (3) forms the neutron beam before the polarizer and increases the effective aperture of the instrument. It is constructed from the glass plates with evaporated nickel coating.

Springer Proceedings in Physics, Vol. 61 **Surface X-Ray and Neutron Scattering**
Eds.: H. Zabel and I.K. Robinson

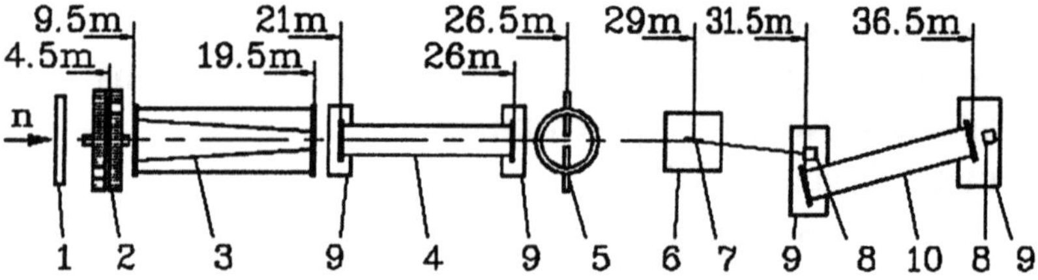

Figure 1. Schematic drawing of the polarized neutron spectrometer at the pulsed reactor IBR-2, Dubna (top view): 1-moderator, 2-double disc chopper, 3-conical neutron guide, 4-polarizer, 5-spin-flipper, 6-turn-table, 7-sample, 8-detector,9-movable platform, 10-analyzer.

The basic spectrometer components are two five meter long bend neutron guides [7], polarizer and analyzer (4 and 10), assembled from iron-cobalt magnetic mirrors containing a titanium-gadolinium antireflecting sublayer. Incident beam polarization is provided using the first neutron guide. The second neutron guide is not being used in the reflectometry mode for the reflected beam polarization analysis. The polarization efficiency of the beam, averaged over the thermal neutron spectrum (the peak of the Maxwellian distribution is at 0.18 nm), is about 95%.

An appropriate static guide field in the polarized incident beam is provided by the permanent magnets with a magnetic field of about 120 Oe. The neutron spin-flipper (5) is a non-adiabatic one [8] reversing the neutron beam polarization over the whole wavelength band with an efficiency of about 99.5% at a beam cross section of $2.5x60\ mm^2$.

The maximum integral neutron flux at the polarizer exit is $1.2x10^6$ n/sec cm^2. However, in the reflectometry mode the strong beam collimation decreases the flux. The horizontal collimation of the incident beam can be varied in the range from $\Delta\theta_{min} = 1.2x10^{-4}$ rad up to $\Delta\theta_{max} = 8x10^{-4}$ rad, and is defined by the cadmium slit at the polarizer exit and the distance equal to 3 meters to the sample. The vertical collimation is $\Delta\theta = 10^{-2}$ rad.

The sample position is located at 29 meters from the source. The sample (7) is fixed in vertical geometry at the turn-table (6). The incident beam angle can be varied in the range from -10^{-2} rad to 10^{-2} rad with an accuracy of 10^{-5} rad. A sample can be settled in the gap of turning magnets. The magnet axes of revolution conforms with the neutron beam direction. The maximum magnetic field value is 10 kOe.

The sample-detector distance is about 2.5 meters. The reflected neutrons are counted by a He^3 detector (8) surrounded by boron-carbide shielding. The detector can be moved across the beam in the range of 150 mm with an accuracy of 0.02 mm. The cadmium slit at the detector entrance is 50 mm in height up to 10 mm in width.

The objective of the specular reflection experiment is to obtain the reflectivity profile as a function of the neutron wavevector transfer normal to the reflecting surface. In the time-of-flight experiment the spectral intensities of the incident beam $N_o(\lambda)$ and of the beam reflected from the surface under investigation $N_r(\lambda)$ are measured at a fixed glancing angle. The method to account the instrumental background is to take measurements of the background spectra $N_{ob}(\lambda)$ of the incident beam and $N_{rb}(\lambda)$ of the reflected beam with the cadmium sheet at the polarizer entrance. Then the experimental reflectivity is calculated by

$$R(\lambda) = (N_r(\lambda)-N_{rb}(\lambda))/(N_o(\lambda)-N_{ob}(\lambda)) \quad (1)$$

where λ is the neutron wavelength. The background conditions provide us a possibility to measure the reflectivity down to 10^{-5}.

In the case of magnetic materials the spin reflectivities R_+ and R_- are measured alternately with the spin-flipper "off" and "on".

3. Experimental Results

The polarized neutron reflection is to be a sensitive probe for the determination of the surface magnetization profile. Much of the interest is now being concentrated on the application of polarized neutron reflection. This is why the production of polarized neutrons remains of great importance. A high polarization of about 95-97% by total reflection from the FeCo film with the TiGd absorbing sublayer for a beam averaged over the thermal neutron spectrum is achieved. However, the measurements of the wavelength dependence of the polarization [9] show an essential decrease of the polarization efficiency for large neutron wavelengths.

To improve the polarizing ability of the magnetic mirrors used in the polarizing neutron guides the magnetic film and the absorbing sublayer were studied separately. The measurements of the reflectivity from Fe50Co50 thin films with the thickness of 150 nm and 110 nm sputtered on the float-glass substrate were carried out for the values of the external magnetic field equal to 300 Oe and 400 Oe. Figure 2(a) shows the reflectivity profiles R_+ and R_- for the 150 nm FeCo thin film in an applied magnetic field of 300 Oe. The neutron scattering length density depth profiles obtained from the fitting of the data are shown in Figure 2(b). To obtain a highly effective polarizing thin film mirror it is necessary to exclude the reflection of neutrons with unwanted spin state, i.e. to obtain $R_-=0$. The investigated magnetic film has not been compensated in the sense that the neutron scattering length density Nb for the unwanted spin state is not zero. The film was not homogeneous in depth

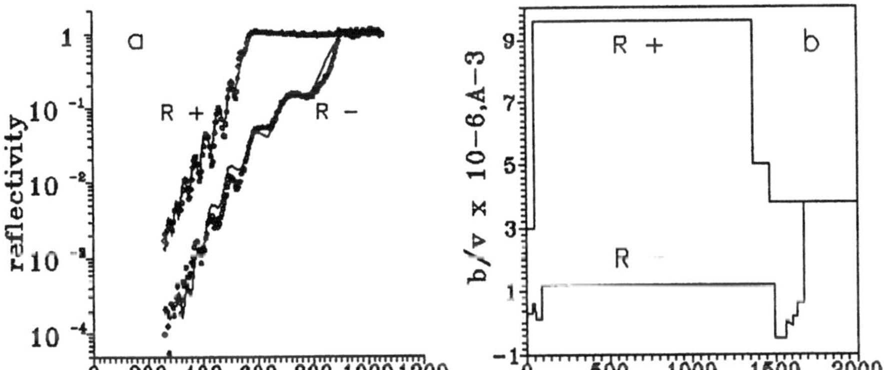

Figure 2. (a) The spin reflectivities for a 150 nm FeCo film on a float-glass substrate in an applied magnetic field of 300 Oe. The solid line is a theoretical fit to the data. (b) The neutron scattering length density profiles obtained from the least-square fit to the experimental data shown in figure 2(a).

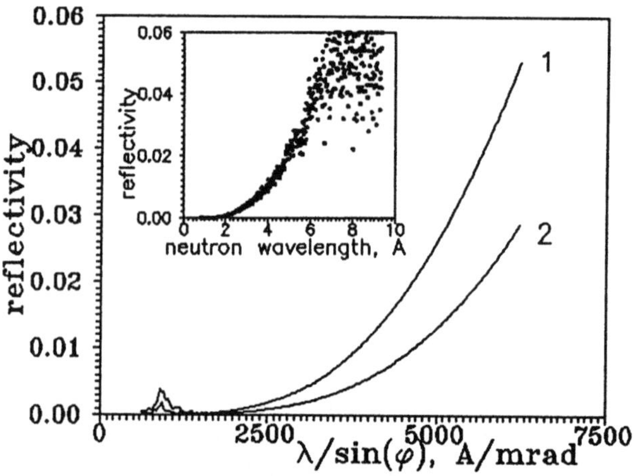

Figure 3. The calculated reflectivity for BTi (curve 1) and BV (curve 2) thin film optimized for the thickness and concentration for the minimum of reflection. The first experimental reflectivity pro file for BTi thin film measured at glancing angle f=5.0 mrad is shown in the insert.

and we assume that the near surface layer was enriched with cobalt. So the problem still remains of further selecting the composition and sputtering conditions for the Iron-Cobalt alloy.

It is worth emphasizing that for the FeCo film the - spin reflectivity (if compared to the + spin one) is very informative, and allows one to find more details of the interfacial neutron scattering length density profile.

The second requirement for the ideal thin film polarizing mirror is that the absorbing sublayer should be non-reflecting for the unwanted spin state. To elucidate the cause of the polarization decrease mentioned above in the large wavelength part of the neutron spectrum ($\lambda>0.4$ nm), the reflectivities from Gd and GdTi thin films were measured [10]. It was shown that a fundamental barrier to obtain a weakly reflecting alloy in a wide region of neutron wavelengths on the base of Gd is the presence of the low-lying resonances in the scattering cross-section which results in a strong dependence of the Gd scattering length on the neutron wavelength. This inconvenience is absent in boron. So one can expect that the boron-based alloys will decrease the reflectivity of the absorbing sublayer and will improve the polarizing ability of magnetic polarizers. Calculated reflectivity profiles optimized for the minimum of reflection for BTi and BV thin films are shown in figure 3. The very first experimental data we have got for a BTi thin film are shown in the insert of figure 3. Now we are on the way to the sputtering optimization of BTi and BV thin films.

The specular reflection of polarized thermal neutrons has been used for the determination of the magnetic field penetration depth Λ into a superconducting single-crystal film of $YBa_2Cu_3O_7$ with T_c=92 K. The film was sputtered on the polished {100} surface of a $SrTiO_3$ single crystal. The sample was cooled in a zero magnetic field to a temperature of 5.2 K. A magnetic field H=200 Oe was then applied in the direction parallel to the surface of the film. A calculation [11] has made it possible to allow for the effect of a nonuniform distribution of the magnetic field in the film and to construct the function $S(\lambda_\perp)=(N-)/(N+)$, which is simultaneously a function of Λ and H, where N_- and N_+ are the measured reflected spectra for two spin components . We assumed that the magnetic induction in the film behaves in accordance with

$$B(z) = H\cosh[(2z-d)/2\Lambda]/\cosh(d/2\Lambda) \quad (2)$$

where z is the distance from the surface along the c axis, d is the film thickness. In an effort to reduce the statistical error in evaluating the effect of a deviation of S from 1, we summed each of the N_+ and N_- spectra over the interval of $\lambda_\perp$ from λ_c to $\lambda_c/2$ (λ_c is the critical neutron wavelength) and the value Λ_c=970(+600/-250) Å was obtained [12].

A further improvement in the statistical base will make it possible not only to determine the parameters Λ_c and Λ_{ab} more accurately but also to define the behavior B(z), by making use of the spectral dependence $S(\lambda_\perp)$.

The experience acquired during the work with the SPN-1 allows us to start at present the work connected with the creation of a specifically new reflectometer of polarized neutrons at the IBR-2 reactor.

References

1. E. Abrahams, W.Zinn: Phys.Rev.70,103 (1946)
2. G.P. Felcher, K.E. Gray, R.T. Kampwirth, M.P. Brodsky: Physica B 136,59(1986)
3. R.Felici, J.Penfold, R.C.Ward, W.C.Williams: Appl.Phys.A 45,169 (1988)
4. B.Farnoux: Proc.Conf. Neutron Scattering in the Ninetics (Viena: IAEA) 1985
5. M.Stamm, G.Reiter, S. Hultenbach: Physica B,156&157,564 (1989)
6. Unpublished results 1983: Neutron research facilities at the Ill high-flux reactor
7. G.M.Drabkin, A.I.Okorokov, A.F.Schebetov, N.V.Borovikova, A.G.Gukasov, A.I.Yegorov, V.V.Runov:JETP 69,1916 (1975)
8. D.A.Korneev, V.A.Kudrjashov:Nucl.Instr.Meth. 179,509 (1981)
9. D.A.Korneev, V.V.Pasyuk, H.Rzany, A.F.Schebetov: JINR Rep. P3-81-546 (1981) and P3-81-547 (1981)
10. D.A.Korneev, V.V.Pasyuk, A.V.Petrenko, H.Jankovski: submitted to Nucl.Instr.Meth.
11. D.A.Korneev, L.P.Chernenko: JINR Rep. R4-87-460 (1987)
12. S.V.Gaponov, E.B.Dokukin, D.A.Korneev, E.B.Kluenkov, W.Loebner, V.V.Pasyuk, A.V.Petrenko, H.Rzany, L.P.Chernenko: Pis'ma Zh.Eksp.Teor.Fiz. 49, No 5, 277 (1989)

Magnetic Properties of Ultrathin Co/Ag Films Investigated by Polarised Neutron Reflection

H.J. Lauter[1], *J.A.C. Bland*[2], *R.D. Bateson*[1], *and A.D. Johnson*[3]

[1]Institut Laue-Langevin, Avenue des Martyrs,
F-38042 Grenoble Cedex, France
[2]Cavendish Laboratory, Madingley Road, Cambridge CB3 0HE, UK
[3]RSRE, St. Andrews Rd., Malvern, Worcs., WR14 3PS, UK

Abstract. We have used polarized neutron reflection to investigate the saturated in-plane magnetization of ultrathin single Co films incorporated into Ag/Co/Ag(001) epitaxial sandwich structures. Strong thickness dependences of the magnetization are observed and for Co thicknesses of 1 and 2ML we find ferromagnetic order at 5K with moments per Co atom close to $2.0\mu_B$, significantly exceeding the bulk hcp phase value of $1.7\mu_B$.

First principles band structure calculations have predicted strongly enhanced magnetic moments in epitaxial Fe and Co metal monolayers supported by noble metals [1,2]. The reduced atomic coordination of the 3d atoms and the large Ag lattice parameter results in a high density of states at the Fermi energy E_F. This mechanism is predicted to be dominant in determining the moment since hybridisation effects between the 3d states and the substrate sp states which reduces the moment are weak for these systems. Cu has a smaller lattice constant which increases the hybridisation for Co/Cu(001). For an unsupported monolayer of Co(001) with the Ag lattice parameter a moment of $2.2\mu_B$ is predicted, falling to $2.03\mu_B$ when epitaxed to Ag(001) [2]. Both values significantly exceed the bulk hcp phase value of $1.7\mu_B$.

Experimental tests of these predictions are demanding, requiring well characterized epitaxial systems and highly sensitive absolute magnetometric techniques, and to date, few experimental tests of absolute moment predictions have been made. We have carried out a polarized neutron reflection (PNR) [3,4] study of the absolute value of the magnetic moment per atom in ultrathin fcc Co single films incorporated into Ag/Co/Ag(001) epitaxial sandwich structures. We find conclusive evidence for enhanced magnetic moments in films of less than 3ML thickness. This observed thickness dependent behaviour of the magnetic moment contrasts with that found for Cu/Co/Cu(001), with moments close to the bulk phase value in the 1-10ML range [3,4].

The experimental validity of the PNR technique as a magnetometric probe of sandwiched ultrathin transition metal films has been established in a series of pioneering experiments [3,4]. The ratio of the spin dependent intensity reflectivities F (flipping ratio) is determined as a function of $q=k\sin\Theta$ corresponding to the perpendicular wavevector of the incident beam. For the symmetric sandwich structure (vacuum(1)/overlayer(2)/magnetic layer(3)/substrate(4)) the peak value of the quantity $F(q)-1$ is directly proportional to the product of the magnetic moment m_3 per atom and the magnetic layer thickness t_3 divided by the atomic volume V_{at} and is enhanced by the overlayer (in which the perpendicular wavevector is q_2) according to [5]:

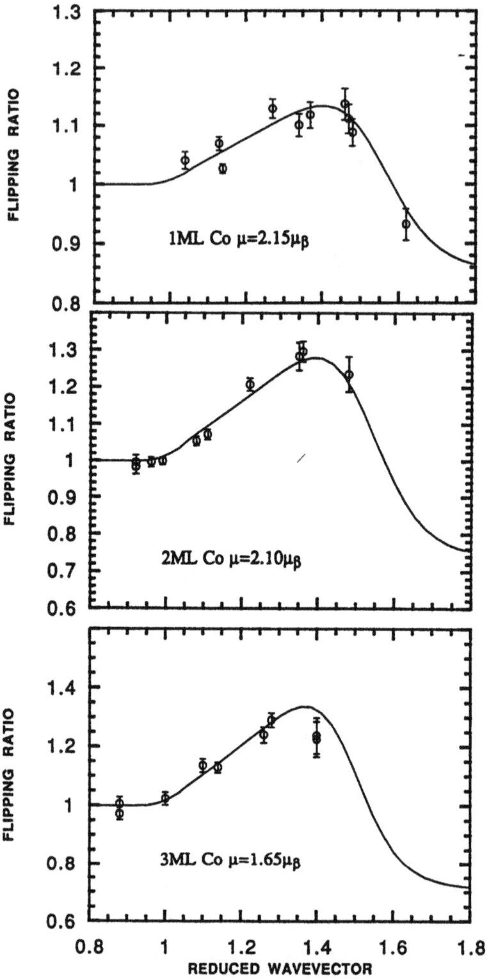

Figure 1: The flipping ratio observed at 5K for a 222Å Ag/1ML Co/222Å Ag film (bottom), a 222Å Ag/2Ml Co/222Å Ag film (middle) and a 230Å Ag/3MLCo/230Å Ag film (top) all on GaAs magnetised in-plane by a 0.83kG field; the solid lines refer to calculations described in the text for the indicated values of the Co moments.

$$(F(q)-1) \sim (m_3t_3/V_{at}) \sin(2q_2t_2) \quad . \tag{1}$$

The PNR technique therefore yields an accurate estimate of μ_{Co} independently of c, the perpendicular lattice parameter, in the ultrathin limit provided the in-plane lattice parameter of the film and its thickness in ML is known.

The films were prepared by evaporation in UHV at 300K on thick epitaxial Ag(001) films deposited on GaAs(001) substrates and also on single crystal Ag(001) surfaces as reported previously [6]. Three well defined breaks in the As-t curve are observed upon completion of successive monolayers. Angle resolved Auger measurements of the growth of Co/Ag(001) have confirmed the stabilization of the

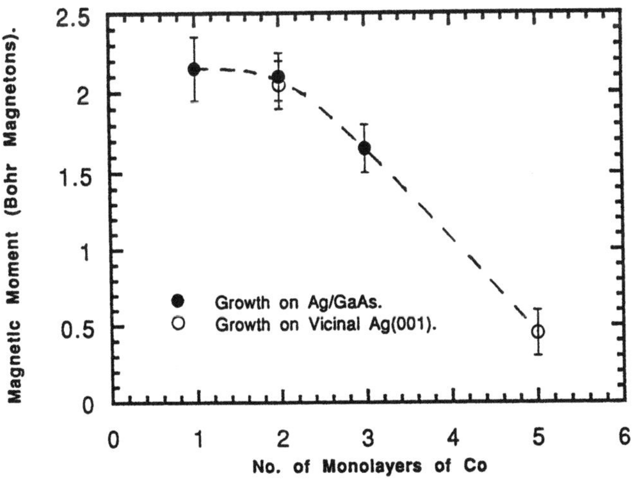

Figure 2: The thickness-dependence of the magnetic moment per Co atom determined by PNR for Ag/Co/Ag (001) structures.

fcc phase for low coverages [7]. This is surprising in view of the 15.4% in-plane expansion required for true epitaxial growth. Beyond 3ML the LEED pattern fades suggesting the high strain cannot be sustained with increasing thickness and that a high density of dislocations and defects occurs.

In figure 1 we show the observed flipping ratio for 1ML, 2ML and 3ML sandwiched Co films. The solid lines in the figures are calculated for the inset magnetic moment and using the the measured film thicknesses with the Co film assumed to be lattice matched in-plane to the Ag substrate. For the 1ML and 2ML films the fitted moment considerably exceeds the bulk value of 1.64μ_B predicted by Li et al [9] for a 5ML fcc Co slab whether we assume a lattice matched fcc structure or the bct structure proposed by Li and Tonner [8] as expected from equation (1). The 2ML thickness samples prepared in single crystal Ag surfaces and on thick epitaxial Ag films yielded the same value of μ_B within experimental error [6]. For a Co thickness of 3ML we find a ferromagnetic moment close to 1.7μ_B with the moment falling significantly below this value for the 5ML film. No increase in the neutron polarization is observed for the 5ML film upon increasing the applied field to 5kG suggesting that the easy magnetization axis lies in plane and perpendicular remanence does not occur.

In figure 2 we summarizes the thickness-dependence of the moment obtained for the film studied. The observed collapse of the 5ML magnetization is attributed to the film being unable to sustain the high substrate-induced strain energy at this thickness and becoming strongly disordered as evidenced by the absence of magnetic anisotropy demonstrated by surface magneto-optic Kerr effect (SMOKE) measurements.

In-plane magnetization was confirmed for all films using (SMOKE) measurements. The angular measurements on the 2ML and 3ML samples reveal large in-plane anisotropies with a strong uniaxial anisotropy in-plane. Th e difference of the saturation fields between the hard and easy axis is 0.7 kOe for the 2ML film and 1.5 kOe for the 3ML film, respectively (see fig.3a for a SMOKE measurement approximately along the in-plane easy axis for the 2ML film). The saturation and

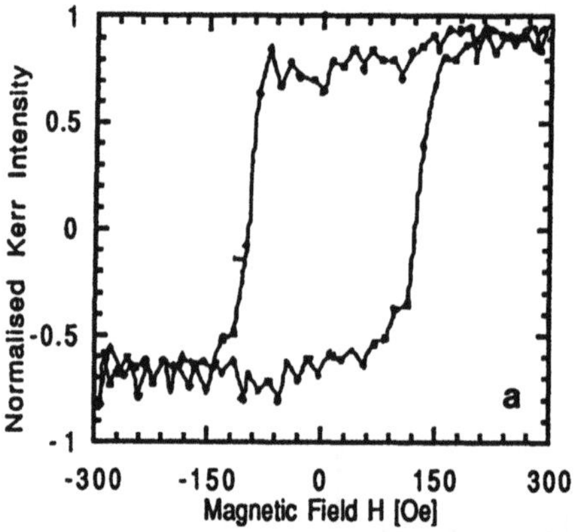

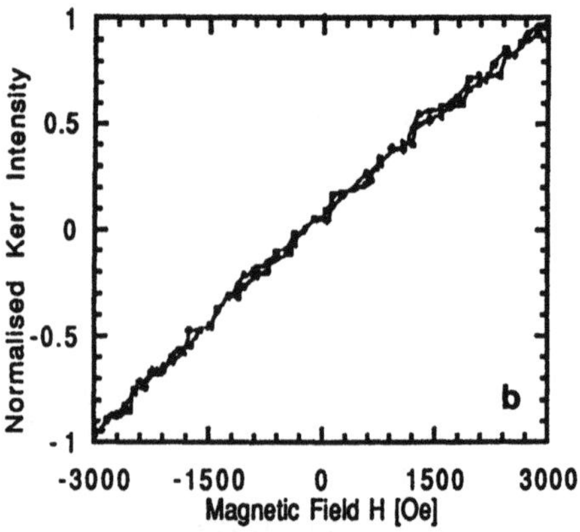

Figure 3: SMOKE measurements of the AG/2ML Co/Ag(100) sample: applied external field parallel (a) and perpendicular to (b) to the film surface.

coercive fields observed for the 3ML film are significantly increased with respect to those observed for the 2ML film, suggesting that the magnetic anisotropy is sensitively dependent on the strain energy and film quality. The film normal is a hard axis along which reversible coherent rotation is observed at low fields indicative of a well defined interface (see fig. 3b).

The interface roughness is negligible for the GaAs supported samples.

In summary we have presented conclusive evidence for strongly enhanced moments in epitaxial single Co/Ag films of less than 3ML thickness. Enhanced moments have previously been reported for Co/Pt and Co/Pd multilayers [10]. The strong thickness dependence of the magnetization and magnetic anisotropy behaviour is shown to be attributable to the high strain in the system, with the collapse of the magnetization and anisotropy occuring beyond 3ML thickness, emphasizing the need for high quality epitaxial systems in order to test theoretical predictions of enhanced moments in ultrathin films.

References

[1] C L Fu, A J Freeman and T Oguchi, Phys. Rev. Lett 54, 2700 (1985); S Blugel, M Weinert and P H Dederichs, Phys. Rev. Lett. 60,1077 (1988)

[2] S Blugel, B Drittler, R Zeller and PH Dederichs, Appl. Phys. A 49, 547 (1989)

[3] D Pescia, R F Willis and J A C Bland, Surf. Sci 189/190 724 (1987)

[4] J A C Bland, D Pescia and R F Willis, Phys. Rev. Lett. 58, 1244 (1987)

[5] G P Felcher, K E Gray, R T Kampwirth, M B Brodsky, Physica 136B,59(1986)

[6] J A C Bland, A D Johnson, H J Lauter, R D Bateson, S J Blundell, C Shackleton and J Penfold, J.Mag. and Mag. Mat. 93 (1991) 513

[7] A D Johnson, PHD dissertation, University of Leicester, 1990

[8] Hong Li and BP Tonner, Phys. Rev. B 40, 10241 (1989)

[9] C Li, A J Freeman and C L Fu, J. Magn. and Magn. Mat. 75, 53 (1988)

[10] C J Lin, G L Gorman, C H Lee, R F C Farrow, E E Marinero, H V Do, H Nortarys and C J.Chien, J.Mag. and Mag. Mat. 93, 194 (1991)

Depth Selective Real Structure Analysis of Semiconductor Superlattices Using Grazing Incidence X-Ray Diffraction

U. Pietsch

Brandenburgische Landeshochschule Potsdam, FB Physik,
Am Neuen Palais 10, O-1571 Potsdam, Fed. Rep. of Germany

Abstract. The periodicity of superlattices(SL) and its deviation perpendicular to the surface can be detected using X-ray grazing incidence diffraction (GID). Depth resolution is obtained by measuring rod scans at angles of incidence smaller or larger than the critical angle of total external reflection. First experimental curves measured from a lattice matched SL are interpreted in terms of a kinematical approach. The capability of the method for a depth resolved characterization is dicussed by simulation of the GID pattern for a partially relaxed strained layer SL.

1. GID measurements from a lattice matched superlattice

Superlattices (SL) consist of a periodic double layer sequence of two different materials and can be characterized by use of conventional X-ray diffraction. Only averaged structure parameters can be determined because the penetration of the X-ray beam is much larger than the total thickness of SL.

Depth resolution is obtained by the technique of grazing incidence X-ray diffraction (GID). This is experimentally demonstrated for a 10 period GaInAs/InP[001] structure measured at the (220) Bragg reflection [1](fig.1). Here the "information depth" is controlled by the angles of incidence α_i and exit α_f of the X-ray beam with respect to the sample surface. The vertical density profile is probed by measuring the intensity distribution along α_f (rod scan) for different α_i at any lateral Bragg angle θ_B which coincides for both sublattices. For α_i smaller than the critical angle of total external reflection α_c (~0.27^o) the information depth is reduced to the top layer and/or the upper SL period. With increasing α_i further SL-cells contribute to the scattering signal and "SL-peaks" appear. From their angular distance the thickness of the supercell can be determined (t_{SL} = 14.0±0.3 nm). The thickness of the solid solution layers t_{ss} within any supercell is obtained in monolayer accuracy (t_{ss} = 0.9±0.3 nm) by making use of the peak splitting which results from the different refraction shift of both sublayers.

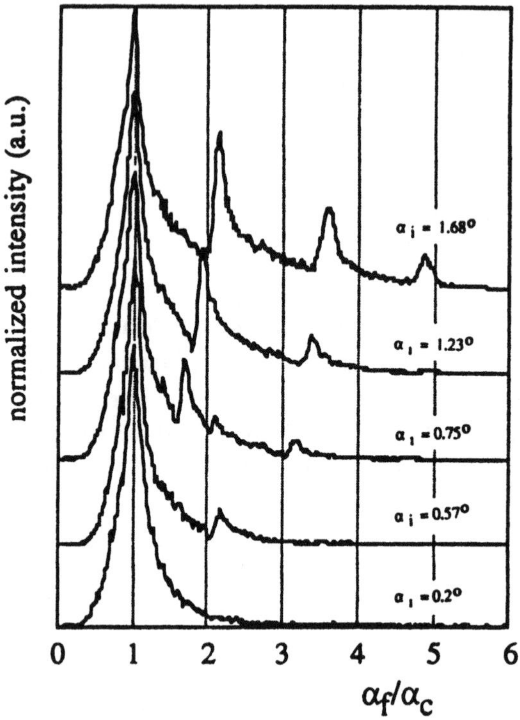

Fig.1: α_f-resolved GID-intensity of the (220) Bragg reflection of a lattice matched $Ga_{0.47}In_{0.53}As$ / InP [001] SL measured for different α_i.

2. Simulation model

The experimental curves can be sucessfully explained applying a kinematical approach of GID[1-3]. If the interfaces between the sublayers are not abrupt on the atomic scale over the irradiated sample area the multiple reflection of the incident and diffracted beam within the SL can be neglected. Then the scattering power $P(\alpha_f)$ is simply described by

$$P(\alpha_f) \sim |T(\alpha_i)\ T(\alpha_f) \int dz\ X_H\ \exp(-i\, q_z z)\,|^2 . \qquad (1)$$

The Fresnel's transmission functions $T(a_{i,f})$ are calculated taking into account an averaged density of SL. The periodicity in depth is solely expressed by the different scattering amplitudes X_H of the sublayers and by changing the scattering vector

$$q_z = q_i - q_f = 2\pi / \lambda\ \{(\alpha_i^2 - X_0(z))^{1/2} + (\alpha_f^2 - X_0(z))^{1/2}\} \qquad (2)$$

from layer to layer. X_0 means the dielectrical susceptibility at q= 0, λ stands for the wave length. The"information depth" is controlled by the imaginary part of q_z. The angular distance between the SL-peaks is proportional to

$$\Delta\alpha_f \sim \lambda / t_{SL} . \qquad (3)$$

3. GID simulation for a partially relaxed superlattice

The capability of the GID-technique for a depth resolved real structure analysis will be demonstrated with the example of a SL with decreasing perfection below the surface. This may be expected in strained layer SL, as for GaInAs/GaAs. If t_{ss} is larger than any critical thickness to create misfit dislocations, the lateral lattice parameters a_{ss} and a_{GaAs} do not further coincide and the lateral Bragg condition can be fulfilled only for one kind of the sublayers. It is known [4] that the relaxation begins at the SL to substrate interface and is gradually extended to the neighbouring strained layers. This effect is simulated in fig.2. which compares the GID pattern of a fully strained and a partially relaxed SL at the Bragg angle of the substrate. The distortion is localized at the three GaInAs layers near the substrate and may be so strong that this layers do not scatter in direction of the fixed Bragg angle. For $\alpha_i = 1.08 * \alpha_c$, the GID signal stems only from the upper regions of sample and the distortion does not influence the rod scan (fig 2a,b). For $\alpha_i = 1.52 * \alpha_c$, however, we find a supermodulation of the rapid oscillations near the SL peaks in the partially relaxed case with a period corresponding to the distance between the "perfect"part of SL and the substrate. Any lateral inhomogeneity of the sample and the mosaic spread of the disturbed layers levels the sharp oscillations and only broadened SL peaks will be detected. Because the distortion increases with depth the SL peak width increases with α_i. Additionally, if $\theta_{ss} \neq \theta_{GaAs}$ the rod scans can be recorded at different Bragg angles. In this case the state of relaxation can be investigated in detail, because the GID-intensity distribution changes its shape in a manner which is characteristic for the layer currently fulfilling the Bragg condition.

The reachable resolution with depth depends on the difference of the optical densities of both sublayers and from the absorption of the material used. In the

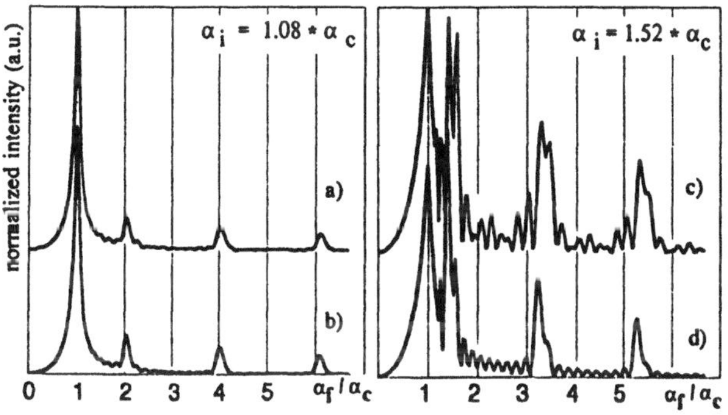

Fig 2: Simulation of α_f-resolved GID intensity of a $Ga_{0.8}In_{0.2}As/GaAs[001]$ strained layer SL (t_{ss}=2nm, t_{GaAs}=12nm) calculated for two different α_i at the (220) reflection of GaAs. b), d) fully strained SL, a), c) partially relaxed state.

present case the "information depth" scans through the total SL-thickness between $\alpha_c < \alpha_i < 2^* \alpha_c$.

Acknowledgements: The author would like to thank the Alexander von Humboldt Foundation for its financial support and Dr.H.Metzger (München) for his permanent assistance during this work.

References:

[1] U.Pietsch, W.Seifert, J.-O.Fornell, H.Rhan, H.Metzger, S.Rugel and J.Peisl Proc. of the E-MRS Spring Meeting ,Straßbourg 1991

[2| G.Wallner, E.Burkel, H.Metzger, J.Peisl and S.Rugel Mat.Res.Symp.Proc. 143 19 (1989)

[3] U.Pietsch, H.Rhan, A.L.Golovin, and Yu.Dmitriev, Semicond.Sci. & Technol. 6 (1991) in press

[4] W.Wegscheiter, K.Eberl, G.Abstreiter, H.Cerva and H.Oppolzer Appl.Phys.Lett. 57 1496 (1990)

Investigation of Interfaces with Grazing Incidence Neutron Radiation

V.I. Mikerov, A.V. Vinogradov, I.V. Kozhevnikov, F.A. Pudonin, V.A. Tukarev, and M.P. Yakovlev

Lebedev Physics Institute, Leninskiy Prospect 53, SU-117924 Moscow, USSR

Abstract. The potential of neutron reflectometry for the investigation of statistical properties of interfaces, large scale magnetic structures and others is demonstrated.

1. Introduction

The development of nondestructive methods of investigation and quantitative control of the surface microstructure is important both for physical research and for modern technology. Methods of particular interest are based on the analysis of the angular scattering distribution of X-rays, visible light and cold neutrons [1-6]. It should be emphasized that only the scattering diagram (but not the specular reflectivity of the total integral scattering) contains the reliable quantitative information about the statistical properties of surface microstructure.

The first results of neutron scattering measurements carried out with the neutron diffractometrical set-up DNM-1 are discussed in this work. Three problems are considered:
- the determination of surface roughness spectra;
- the study of magnetic scattering;
- the interface structures in multilayer systems.

2. Experimental conditions

The measurements were carried out with the neutron reflectometer being a part of the set-up DNM-1 situated at the steady state reactor of the Moscow Engineering Physics Institute (2.5 MW). The wave length range was from $\lambda = 2.15$ Å to $\lambda = 6.59$ Å with a spectral resolution of less than $\Delta\lambda < 0.05\,\lambda$. The angular divergence of the neutron beam was equal 1.5′. This value was determined by a two-slit collimator. The angular resolution of the detector was 0.7′.

3. The determination of surface roughness spectrum

As shown in [2,3] it is possible to determine the spectrum of surface roughnesses $\chi_c(p)$ from the scattering diagram directly in some range

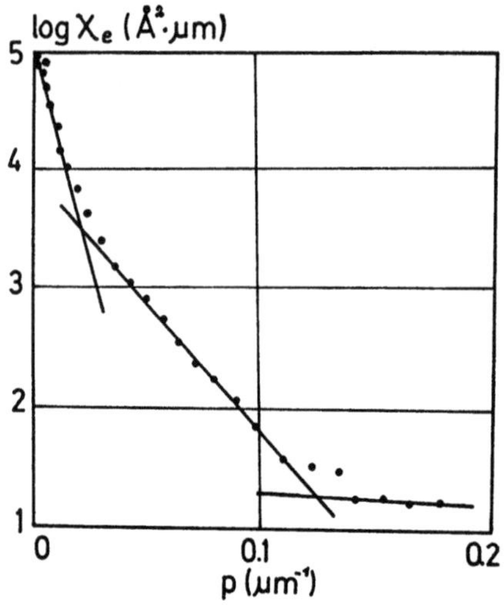

Fig. 1: The roughness spectrum of external surface of 1000 Å nickel film.

of spatial frequency values p. The function $\chi_c(p)$ is the cosine Fourier transform of the correlation function $\chi(\rho)$ of the surface roughness heights.

The experimental roughness spectrum $\chi_c(p)$ of an external surface of a 1000 Å thick nickel film evaporated onto floatglass substrate is shown in Fig. 1. The solid lines are the theoretical fits under assumption of exponential dependence of $\chi_c(p)$. Therefore it may be concluded that three independent roughness microstructures with different roughness heights ζ_j and correlation lengths a_j exist on the surface:

$$\chi_c(p)=\sum_{j=1}^{3} \sqrt{\pi/2}\ \zeta_j^2\ a_j\ \exp(-p\ a_j)\ , \qquad (1)$$

where

$a_1 = 180$ μm; $a_2 = 50$ μm; $a_3 = 2.8$ μm
$\zeta_1 = 32.5$ Å ; $\zeta_2 = 11.7$ Å ; $\zeta_3 = 2.8$ Å .

It means that the correlation function $\chi_c(p)$ is the sum of three functions:

$$\chi(\rho)=\sum_{j=1}^{3} \zeta_j^2\ (1+\rho^2/a_j^2)^{-1}\ . \qquad (2)$$

The above consideration shows clearly that the scattering diagram permits to determine quantitatively the statistical characteristics of supersmooth surfaces.

4. The study of magnetic scattering

The magnetic scattering was first observed by the authors of [7]. The experiments presented in this work are made with iron film (thickness is about 1 μm) evaporated onto a superpolished glass substrate. In Fig. 2, 3 the diagrams of the neutron scattering from a demagnetized sample and a sample placed into a constant magnetic field with strength H = 600 G are shown. The scattering angles refer to the incident beam direction.

The analysis of the scattering diagrams permits to make the following conclusions:

- Considerable magnetic scattering caused by the domain structure of the iron film is present (Fig. 2). The angular width of scattering peaks corresponds to about one-micrometer sizes of domains.
- The additional (Ioneda) peaks of scattering are clearly observed far from the specular reflected beam. As shown theoretically in [1,2] the angular positions of these peaks are determined by the critical angle of the total external reflection. In neutron experiments this effect is observed first.

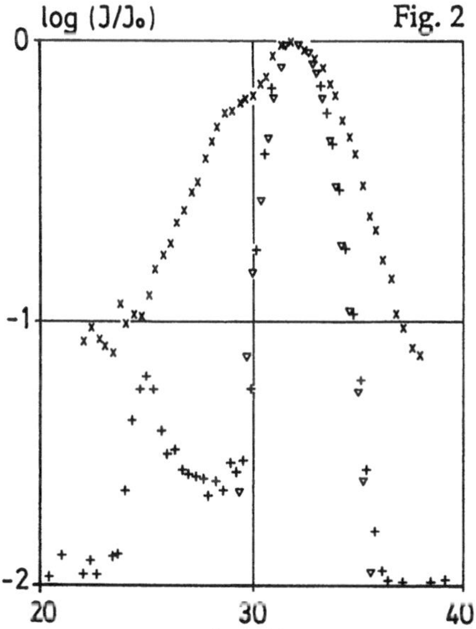

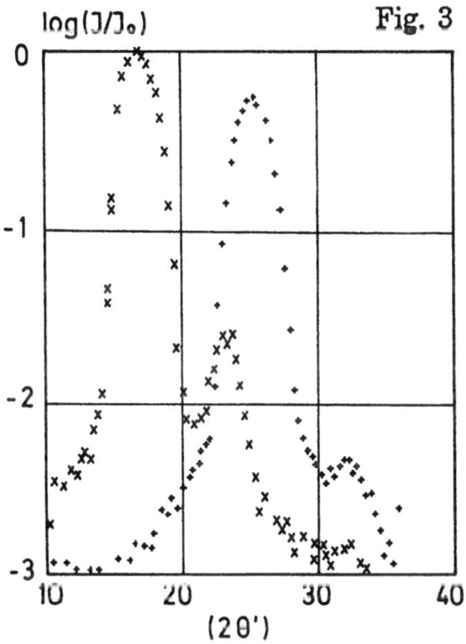

Fig. 2: Scattering diagrams for nonmagnetized (x) and magnetized (+) iron films (1μm) at λ= 2.33 Å. The primary beam intensity distribution is indicated by (∇). The grazing angle was equal to the second critical angle.

Fig. 3: Scattering diagrams for magnetized iron film (1μm) at λ= 2.33 Å. (x) and λ= 2.99 (+) (grazing angle in both cases was equal to the first critical angles)

- The values of nuclear and magnetic potentials are found from the angular positions of the additional peaks of scattering. As it turned out these potentials are about 10 % less than the corresponding values for iron. This difference is probably caused by smaller density of the film in comparison with that of solid.
- Experimental results give the foundation to believe that the neutron depolarization takes place at the scattering process.

5. The interface structure in multilayer systems

One of the main questions in multilayer systems production is conformity of interface microstructure. In the case of some correlation in different interface structures, interference fringes in scattering diagrams should be present. This effect was observed with a Fabry-Perot neutron interferometer. The interferometer consists of two multilayer mirrors separated by 90 Å titanium film. Each mirror consists of 3 pairs of Ni-Ti layers of about 180 Å thickness. Interference fringes are on the right wings of diffuse scattering diagrams (Fig. 4) at the angles of constructive interference for $\lambda = 2.33$

Fig. 4

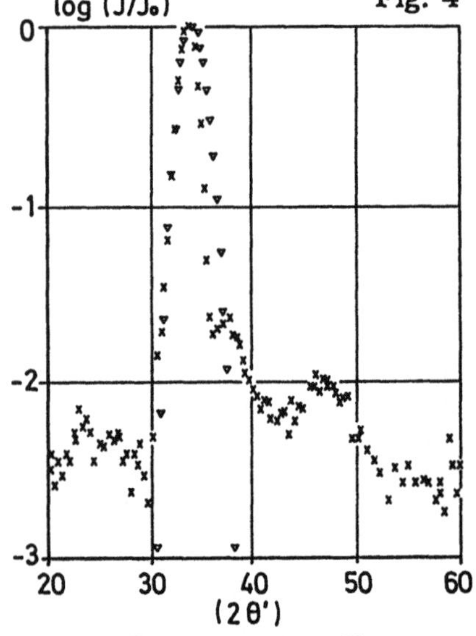

Fig. 5

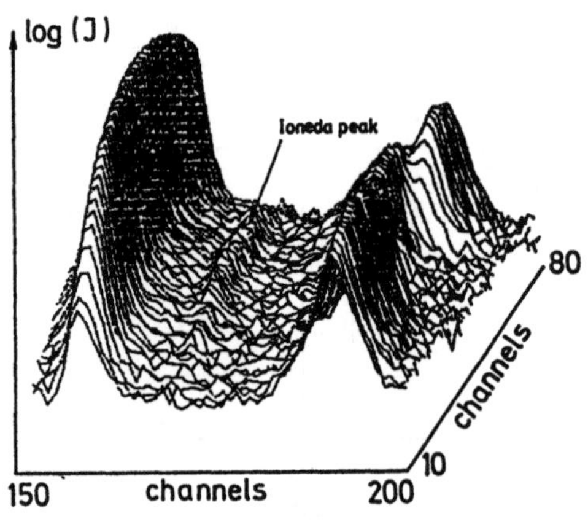

Fig. 4: The scattering diagrams for the Fabry-Perot interferometer at $\lambda = 2.33$ Å. The primary beam intensity distribution is indicated by (▽).

Fig. 5: Neutron intensity distribution measured by a 2-dimensional detector after transmission (right) and reflection (left), respectively, on a Cu-Ti mulilayer structure. The Ioneda peak is seen near the reflected beam.

Å ($2\theta = 38'$). An attempt to produce the same type of a interferometer on the basis of Cu-Ti mirrors failed. Strong diffuse scattering was observed on this sample which resulted in the Ioneda peak. In Fig. 5 one can see the neutron intensity distribution in neutron beams transmitted (right) and reflected (left) by the substrate ($\lambda = 2.33$ Å, $\theta = 15'$); the small peak near the reflected beam is the Ioneda peak.

6. Conclusions

The neutron scattering is a quantitative and high sensitive method of investigations of surface structure of thin films and solids. The use of neutron scattering for investigation of magnetic (domain) structure of surfaces and thin films gives quantitative information about material phase structures. The interference of neutron beams scattered by interfaces must be taken into account by multilayer systems fabrication for X-ray and neutron optics. Magnetic inhomogeneities or a set of interfaces can be used to increase the Ioneda peak intensity and for critical angle measurements.

References

[1] S.K. Sinha et al.Phys. Ref., B 38 (1988) 2297.
[2] A.V. Vinogradov et al. Proc. SPIE, 1140 (1989) 490.
[3] S.S. Borisova et al. Optical & Acoustical Review, 1 (1990) 188.
[4] A. Steyerl. Z. Physik. 254 (1972) 169.
[6] A.V. Antonov et al. JETP Lett., 44 (1986) 448.
[6] L. Nevot and P. Croce. Rev. de. Phys. Appl., 15 (1980) 761.
[7] A. V. Antonov et al. Proc. Ist Intern. Conf. on Neutron Physics (Kiev, September 1987).

Roughness Characterization of the Surface and Interface of MBE-Grown Thin Films

A. Stierle, A. Abromeit, K. Bröhl, N. Metoki, and H. Zabel

Ruhr Universität Bochum, Experimentalphysik IV,
Universitätsstr. 150, W-4630 Bochum 1, Fed. Rep. of Germany

Abstract. Co(0002) oriented thin films were grown by MBE techniques on $Al_2O_3(11\bar{2}0)$ oriented substrates with a typical miscut of 0.5° – 1°. The samples were characterized by high resolution x-ray reflectivity measurements and Laue oscillations of the (0002) Bragg peak, yielding a density profile for the film including a surface oxide layer and interfacial roughness parameters.

1. Introduction

Co thin films have recently attracted much interest because of their intriguing structural and magnetic properties. With a proper choice of the substrate the metastable fcc and bcc phases may be stabilized at room temperature promising new magnetic properties. In addition, the spin anisotropy changes from perpendicular to in-plane beyond a film thickness of five monolayers. For the investigation of the physical properties of the magnetic thin films, careful sample preparation and structural characterization is vital. We have analyzed our MBE grown Co films in-situ by RHEED techniques and ex-situ by x-ray reflectivity and Laue oscillations of the Bragg peak providing information on the film thickness and the number of coherently scattering lattice planes together with their roughness parameters.

2. Experimental

The Co films were grown in a RIBER EVA 32 MBE machine with a base pressure better than 5×10^{-11} Torr, not exceeding 5×10^{-10} Torr during growth. The sapphire substrates were pre-cleaned using an acetone solution followed by an ion milling process in the loading chamber with a 600eV Ar gun. In the main chamber the substrates were heated to 1100°C. At 1000°C a large pulse in the mass spectrometer indicated that the Ar was leaving the surface. In addition, at the maximum annealing temperature a RHEED pattern was observed characteristic for the reconstruction of the $Al_2O_3(11\bar{2}0)$ surface [1]. Co was evaporated by a 10kV e-beam with a growth rate of 0.5Å /s and substrate temperatures between 350°C and 450°C. Subsequently the film was annealed for 30 min at 500°C and 550°C. Best sample quality was achieved at a substrate temperature of 450°C yielding rocking widths of less than 0.04°. The Co films were characterized by high resolution x-ray reflectivity measurements and Bragg scans using $MoK_{\alpha 1}$ radiation and a Si (220) monochromator providing a resolution of $\Delta K/K = 10^{-4}$.

3. Total reflectivity measurements

X-ray reflectivity measurements of thin films deposited on substrates yield information on the electron density profile normal to the film plane. This includes the interfaces between the vacuum and an oxide layer, between the oxide layer and the metal film, and finally between the metal film and the substrate. For the different interfaces the boundary conditions for the reflectivity of the electromagnetic waves have to be fulfilled. This leads to a recursive formalism for the calculation of the reflected intensity, first introduced by Parratt [2]. In the present model the absorption of all layers has been included, which is not negligible even for *thin* Co films, as well as a term accounting for the roughness of the interfaces. Assuming that the interface heights at different lateral positions exhibit a Gaussian distribution, Nevot and Croce [3] deduced a modified Fresnel reflectivity given by

$$R = R_{F_{n,n-1}} e^{-8\pi^2 K_n K_{n-1} \sigma_n^2},$$

where $R_{F_{n,n-1}}$ is the Fresnel reflectivity of the interface n, K_n and K_{n-1} are the wave vectors in medium n-1 and n , σ is the root mean square of the interfacial height, interpreted as the *roughness* of the interface. Fig. 1 shows a typical reflectivity measurement from Co on sapphire. The

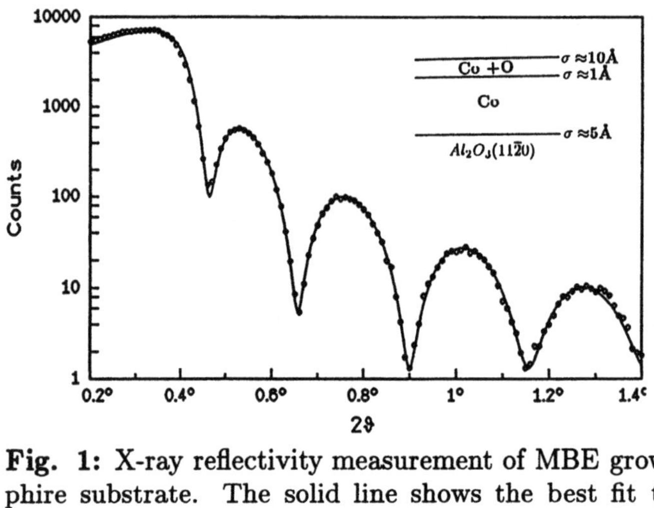

Fig. 1: X-ray reflectivity measurement of MBE grown Co film on sapphire substrate. The solid line shows the best fit to the data points ($\chi^2 = 7 \times 10^{-3}$) with a total film thickness of 180 Å. The inset shows the determined roughness parameters for the interfaces.

solid line represents a fit to the data points using the above described model. The fitting parameters are shown in the inset of Fig. 1 and the resulting electron density profile is reproduced in Fig. 3 a. An oxide layer of 20 Å with a roughness of 10 Å on top of the pure Co part can be deduced. The roughness of the interface Co/Al_2O_3 is determined to be 5 Å. The latter value is consistent with the RHEED observation of a pseudomorphic growth for the first few Co layers.

4. Laue oscillation of the Bragg peak

The Bragg reflected intensity of a finite number of lattice planes with an interplanar separation d is given by the well known Laue function. In our model we assume a Gaussian distribution of the number of coherently scattering lattice planes. The Gaussian superposition of the Laue functions is carried out incoherently, leading to the following result:

$$I \sim \sum_N \frac{sin^2(\frac{NK_\perp d}{2})}{sin^2(\frac{K_\perp d}{2})} \cdot \frac{1}{\sqrt{\pi}\sigma} e^{-\frac{(N-N_0)^2}{\sigma^2}},$$

where N_0 is the average number of lattice planes, $K_\perp$ is the scattering vector normal to the surface, and σ is the width of the Gaussian distribution. Here the parameter σ is a measure of the roughness of the coherently scattering planes. Fig. 2 reproduces plotted a Co(0002) Bragg peak exhibiting the typical Laue oscillations. According to the fit to the data points (solid line), the average number of coherently scattering planes is 71, which corresponds to a film thickness of 145 Å. This compares

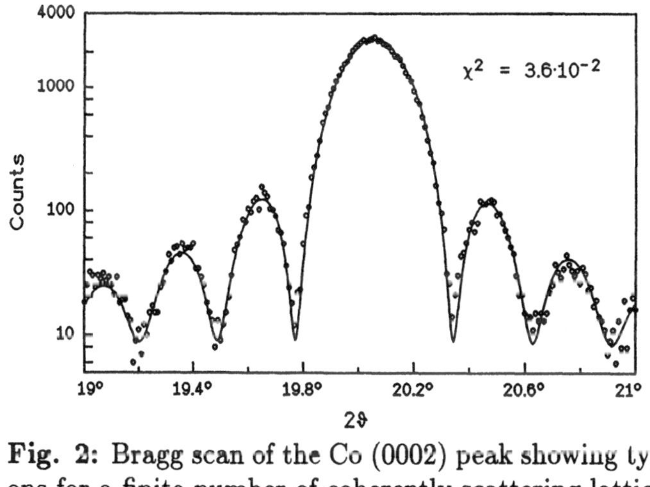

Fig. 2: Bragg scan of the Co (0002) peak showing typical Laue oscillations for a finite number of coherently scattering lattice planes. The solid line represents a best fit to the data points with an average of 71 layers (≅ 145 Å) of coherently scattering planes.

well with the total Co film thickness of 149 Å as determined from the reflectivity data.

5. Discussion and Conclusion

Fig. 3 shows the electron density profiles derived from the total reflectivity measurement and from the Laue oscillation. From the electron density profile we find an oxide layer on top of the pure Co film with a thickness of 20 Å. This thickness agrees well with the saturation thickness of CoO on Co as deduced from magnetization measurements [4]. The roughness of this layer is about 10 Å. The most interesting result is the fact that the thickness of the pure Co region determined by total reflectivity is only a little bit larger than the stack of coherently scattering lattice planes. This indicates that the metal film is highly correlated normal to the surface. The roughness of 7 Å determined from the damping of the Laue oscillation is in good agreement with the interfacial roughness of 5 Å determined from the total reflectivity measurement.

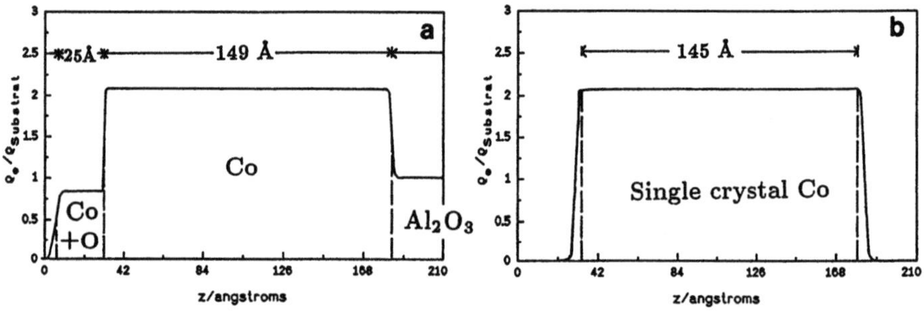

Fig. 3: Electron density profile normal to the film plane. (a) density profile as determined from the reflectivity measurment, and (b) from the Laue oscillation. Note that in both profiles the Co thickness is roughly the same, indicating that the Co film is well ordered over the total film thickness.

This work was supported by the Deutsche Forschungsgemeinschaft (SFB 166), and by the Ministerium für Wissenschaft und Forschung NRW, which is gratefully acknowledged.

References

[1] C.C.Chang, J. Appl. Phys. **33**, 5570 (1968)
[2] L.G. Parratt, Phys. Rev.**95**, 359 (1954).
[3] L. Nevot and P. Croce, Revue Phys. Appl.**15**, 761 (1980).
[4] U. Köbler, L. Smardz, and W. Zinn, Verh. DPG, p. 1222 (1991).

Part VII

Instrumentation and Methods

Neutron Diffraction Under Grazing Incidence: Recent Results from the Evanescent Wave Diffractometer

K. Al Usta, H. Dosch, A. Lied*, and J. Peisl*

Sektion Physik der Universität München,
W-8000 München, Fed. Rep. of Germany
*Present address: Institute Laue-Langevin, P.O. Box 156X,
F-38042 Grenoble Cedex, France

Abstract: We report the first exit angle resolved magnetic neutron diffraction experiment under grazing incidence. The antiferromagnetic (100) Bragg reflection was excited under grazing incidence on a (001) MnF_2 single crystal surface. By a careful control of the exit angle we were able to identify unambiguously the surface scattering. We present the experimental set up and compare some results with those of a CaF_2 single crystal.

1. Introduction

Since Marra and Eisenberger [1] have reported the first surface sensitive X-ray diffraction experiment, the technique has rapidly developed and is today a well established technique in surface science [2]. The development of a grazing angle neutron scattering technique together with the construction of a suitable instrumentation .have been so far neglected, despite the obvious advantage neutrons would have for the study of surface phenomena which are related to magnetism. This circumstances is predominantly due to the very weak surface scattering cross section and the extreme scattering geometry, which both ask for high flux and brilliance in the incident beam. The prospect, however, to obtain information about microscopic magnetic long range order across an interfaces or thin films by applying evanescent neutron scattering has challenged us to develop and test a prototype instrument. The evanescent wave diffractometer (EVA) is placed at the high flux reactor of the Institute Laue-Langevin in Grenoble (France) and is specially dedicated to depth-controlled grazing angle experiments where the scattering is observed out of the plane of incidence.

We have recently reported first exit angle resolved evanescent neutron scattering experiments on CaF_2 and InP single crystal surfaces [3], which demonstrated the feasibility of grazing angle neutron scattering but which made also clear that for neutrons a well defined surface sensitivity is only obtained, when both of the grazing angles are carefully controlled. The data presented now are taken from an recent experiment were we applied this method for the first time to study the surface induced decay of antiferromagnetic ordering near the Neel temperature on a MnF_2 single crystal. This application of grazing angle neutron scattering was first proposed by Dietrich and Wagner [4] who also calculated the corresponding elastic cross section for critical surface scattering. In the following we describe the instrumental setup and compare some of the results with our earlier measurements.

2.The EVAnescent wave diffractometer

Fig.1 shows the schematic scattering geometry that has to be realized in grazing angles diffraction experiments. The surface sensitivity associated with this geometry, is given by the so-called scattering depth Λ, [5] and is controlled by the external angles α_i and α_f. Figure 2 shows the scattering depth for various samples and combinations of these angles. Values of Λ between 100 and 1000 Å are possible if both angles are kept below the critical angle, α_c, for total reflection (α_c~1-2 mrad/Å wavelength). The experiments therefore require good angular resolution in α_i and α_f, and a flat sample surface so that the glancing angles are well defined.

The diffractometer EVA (fig.3) is placed on the guide H53 at the ILL which is supplied with neutrons from the new cold source. Two pyrolytic graphite crystals with intrinsic mosaic spreads η~0.5° are approximately 1° misaligned to each other, and select a small band of wavelengths ($\Delta\lambda/\lambda$~0.01) around λ=5.50 Å out of the neutron guide spectrum. The choice of λ=5.5 Å is a compromise between two conflicting factors: while scattering intensity is gained for large values of λ (since α_c~λ, the vertical divergence can be more relaxed at higher λ), the maximum available horizontal momentum transfer then gets accordingly limited ($|Q_{//max}|$ ~ $1/\lambda$). Our choice of λ, gives $|Q_{//max}|$ ~ 2Å^{-1} and feasible critical angles around 5 mrad. Higher-order wavelengths are suppressed by a liquid nitrogen cooled Be-filter mounted in the flight path between two Cd slits (C0 and C2) which define the beamsize (at the sample position: typically 50*0.4 mm^2) and the α_i resolution. With apertures of 0.5 and 0.2 mm for C0 and C2 the vertical divergence of ~1.6°

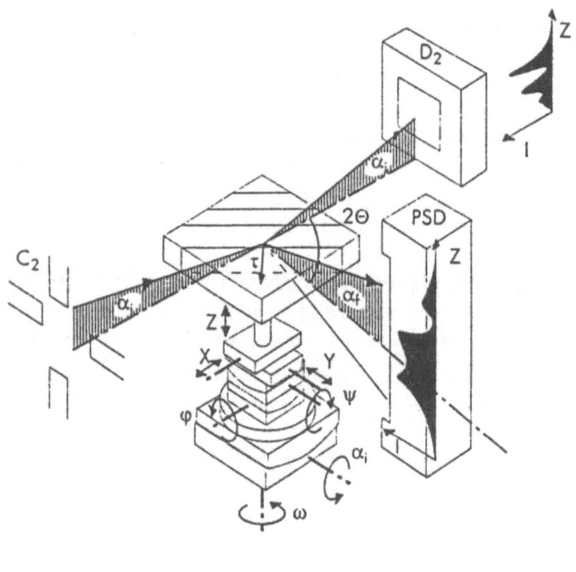

Fig. 1: Schematic view of the sample mounting and the grazing angle scattering geometry.

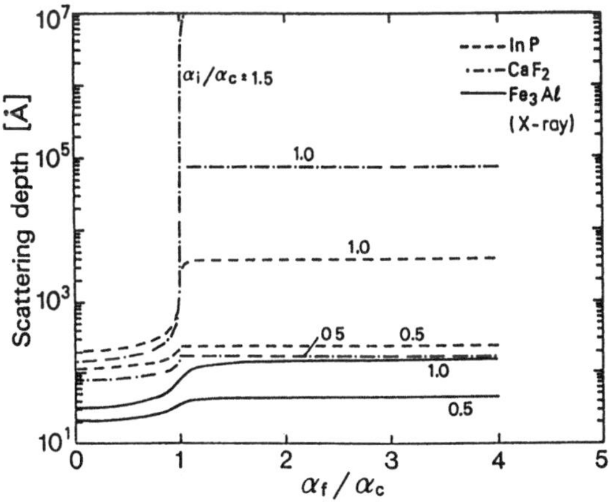

Fig. 2: Calculated scattering depth profiles as a function of the exit angle for various incident angles and samples.

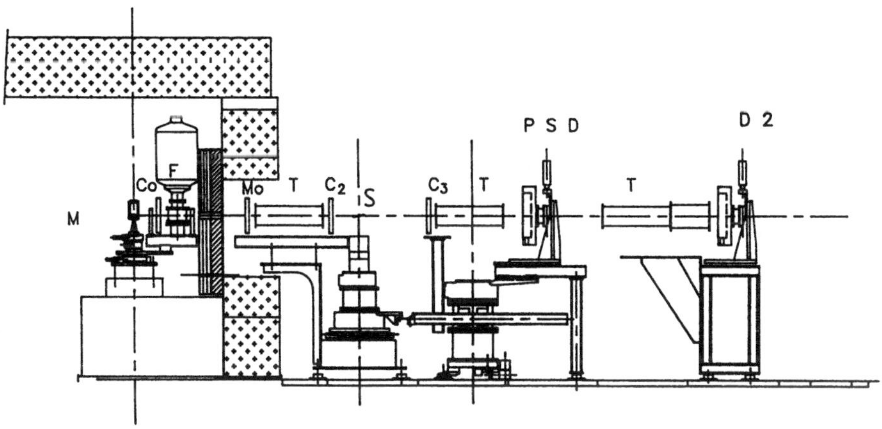

Fig. 3: Schematic side view of the EVAnescent diffractometer (Tanzboden type). M=Monochromator, F=Be-filter, C0,2,3=slits, Mo=Monitor, S=sample position, T= evacuated flight tubes, PSD= position sensitive detector, D2=specular beam detector.

after the monochromator is reduced to an angle uncertainty of $\Delta\alpha_i$~0.4 mrad for the incident neutron beam. The in-plane divergence is in general more relaxed (maximal: $\Delta\omega$~2.5°). The actual flux at the sample position is ~ 2.10^5 neutrons/cm^2 mrad sec. A monitor is mounted before C2 and used to correct the data for intensity variations in the primary beam.

Consider again fig.1 which shows the mechanical setup at the sample position on an enlarged scale: The sample equipment (furnace, cryostat etc.) is mounted on a stepping motor controlled unit consisting of x-y-z translation stages (Δx, Δy, Δz=0.01 mm) and a Φ-χ cradle ($\Delta\Phi$, $\Delta\chi$=0.005°) which is used to align the surface parallel to the incident beam. The angle α_i between the sample surface and the primary beam can be changed with a precision of 0.01 mrad by moving the whole setup on a large cradle, while its absolute value is conveniently determined from the distance between primary and specular reflected beam in the forward direction. For precise measurements of the position and width of the specular peak a ^3He detector (50 mm diam. and 100 mm height) with variable aperture is moved vertically or replaced by a linear position sensitive detector (PSD). The distance between sample and detector is ~1m. Both detectors are surrounded by boron-carbide, cadmium and polyethylene shielding material. The instrumental background with the reactor running and primary beam open, is about 2 cts./min..

Grazing angle diffraction measurements also require sample rotation and a second detector axis to install the in-plane scattering condition (e.g. for Bragg scattering: $\underline{Q}_{//} = \underline{\tau}$, where τ is a reciprocal scattering vector in the surface). The ω table is mounted between α_i cradle and z table so that the total reflection condition is fully preserved while the sample is rotated (ω=0-360°) around the surface normal. For a fixed position of the in-plane scattering angle 2Θ (e.g. $2\Theta_B \approx 2 * \sin(\tau * \lambda / 4\pi)^{-1}$ = 121.4° for the (111) reflection of CaF_2), the evanescent wave scattering is recorded by the PSD which provides the detailed scattering profile along the grazing exit angle. The spatial resolution of the PSD is roughly 1mm (200 pixels with 50 mm height), while the FWHM resolution is about 2 mm and currently limits the α_f-resolution to ~1 mrad for a sample to detector distance of 2 m. The scattering angle 2Θ can be continuously changed between 0 and 135° while its absolute value is controlled by an encoder (resolution $\Delta 2\Theta < 0.01°$) in the sample axis. Sample and detector scans are of considerable interest for critical scattering experiments where the critical diffuse scattering becomes more and more important when the transition temperature is approached.. By scanning these axis the diffuse scattering can in general be separated from the long-range order contribution.

3. Experimental results

Using the instrument described above, we have so far carried out GID Bragg scattering experiments from Si, CaF_2, InP and ,recently, MnF_2 single crystals. Grazing angle neutron experiments on Si have already been carried out before by Zeilinger and Beatty [6], who used a somewhat simpler but less surface sensitive asymmetric scattering geometry to measure the (111) Bragg reflection from a (211) Si surface. Another GID experiment on a Si (110) surface was carried out by Ankner et al.[7] who measured the (111) Bragg reflection under grazing incidence in the geometry of fig.1. However, the surface sensitivity which was obtained in their experiment was approximately the same as in the experiment by Zeilinger et. al, because they integrated experimentally over the whole range of α_f angles. Besides the lower depth resolution of such experiments one has also to worry about bulk scattering which may also be counted by an open detector since neutrons are, as we will demonstrate below, particularly unforgiving when the exit angle is relaxed.

We begin with the measurement on CaF_2: Fig.4 shows the α_f -resolved Bragg profile of the nuclear (111) CaF_2 reflection, as obtained for an angle $\alpha_i \approx \alpha_c$ (α_c=6.2 mrad) between the ($\underline{2}$11) sample surface and the incident beam (the beam divergence was $\Delta\alpha_i/\alpha_c \approx 0.1$). We find three different contributions to the Bragg scattered intensity which are scattered at distinctly different exit angles. The peak at $\alpha_f \approx \alpha_c$ ("S" in fig. 4) belongs, as in the X-ray case to the pure surface diffracted beam. Its

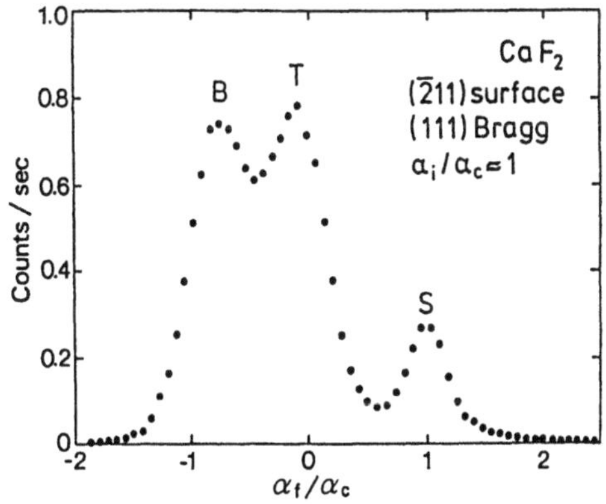

Fig. 4: α_f-profile of the (111) Bragg reflection from a CaF_2 ($\bar{2}$11) surface observed under a grazing incident angle of $\alpha_i/\alpha_c \approx 1$. For an explanation of B,T and S see main text.

α_f position was found to be independent of α_i as expected from the distorted wave Born approximation (DWBA). Note that only 1 count / 5 s has been scattered into the surface peak. At $\alpha_f \approx 0$ the transmitted diffracted intensity ("T") is found which originates from neutrons impinging onto the sample surface and leaving the crystal through the exit face. While this beam has nevertheless a moderate surface sensitivity, the third peak ("B") belongs to ordinary bulk scattering where the neutrons enter and leave the crystal through side faces and which contains no surface information at all. These contributions are visible due to the small neutron absorption cross section of CaF_2, which results in a large absorption length of la≈ 31 cm. For further experimental details and a comparison with evanescent scattering from material with lower absorption lengths, as InP (la=0.9mm), we refer the reader to [3].

In the second example we show preliminary data obtained from a MnF_2 single crystal with the surface oriented parallel to the (001) lattice planes. MnF_2 is a classical antiferromagnet with a Neel temperature of $T_N \approx 67.4$ °K [8]. Below T_N the spins of the Mn^{2+} ions are aligned parallel to the tetragonal c axis with a spin structure which gives rise to a pure magnetic (100) Bragg reflection [9]. We have studied the temperature dependence of the magnetic (100) reflection in a GID experiment, in order to measure the surface induced decay of antiferromagnetic ordering [10]. Fig 5 shows a typical α_f-profile from MnF_2, for an incident angle $\alpha_i \approx \alpha_c$ (α_c=4.28 mrad, the beam had $\Delta\alpha_i/\alpha_c \approx 0.17$) at a temperature T≈61.7K. Although the absorption length of MnF_2 (la ≈7.5 mm) is smaller than for CaF_2 we find again the same scattering contribution of bulk, transmitted diffracted and surface diffracted intensity present. Again only 1 count / 5 s is scattered in the magnetic surface peak ("S). Note also that the position of the transmitted and bulk diffracted beams are shifted to lower α_f values compared to the corresponding peaks in fig4., while the maximum position of the surface scattered intensity is remained unchanged at $\alpha_f \approx \alpha_c$. This shift is due to a small miscut of ≈ 6 mrad between the surface and the (001) lattice planes.

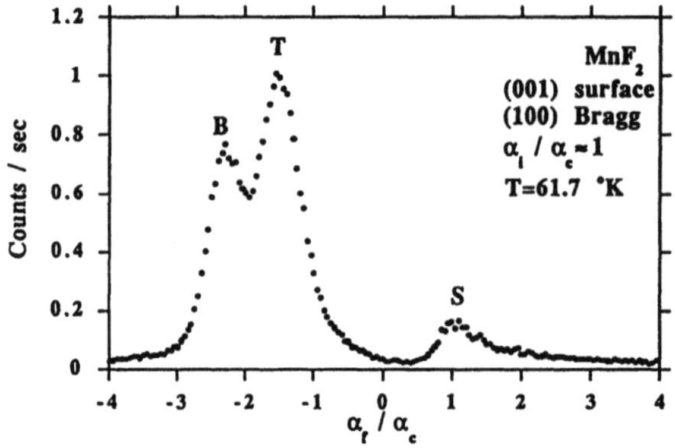

Fig. 5: α_f-profile of the pure antiferromagnetic (100) Bragg reflection from a MnF_2 (001) surface as measured at a temperature of T≈61.7 °K .The angle of incidence was again $\alpha_i/\alpha_c \approx 1$.

4. Summary and conclusion

We have performed the first magnetic (exit angle resolved) GID experiment on an antiferromagnetic crystal. We have demonstrated how grazing angle neutron diffraction experiments have to be carried out, to separate scattering which originates from the surface from bulk scattering. This was most conveniently achieved by the use of a position sensitive detector. Our experiments show that in a neutron GID experiment, contrary to the X-ray case, the surface scattered intensity of a magnetic reflection is comparable to that of a nuclear reflection. The example also demonstrates the advantage of neutron scattering under grazing incidence against magnetic neutron reflection, since an antiferromagnetic ordering is undetectable in the specular direction.
We are currently continuing our experiments on critical magnetic surface scattering. From the data presented here one has to conclude ,however, that the application of this method is limited to the investigation of long range order structures and that the proposed measurement of near-surface critical diffuse scattering appears rather unrealistic. Although the possibility to measure bulk and surface critical behavior simultaneously seem intriguing, the diverging scattering cross-section for the critical bulk fluctuations may cause a general background problem in such experiments.

Acknowledgements

We thank the Institute Laue-Langevin for the support and the opportunity to develop and run the instrument EVA. We are greatly indebted to Dr. W. Drexel, Ph. Ledebt, J.L. Coquin and N. Givelet for their kind assistance. This work is supported by the Bundesminister für Forschung und Technik under Contract No. 03PE1LMU2.

References

[1] W. C. Marra, P. Eisenberger, and A. Y. Cho, J. Appl. Phys. **50**, 6927 (1979)

[2] R. Feidenhans'l, Surf. Sci. Rep. **10**, 3 (1989); H. Dosch, L. Mailänder, A. Lied, J. Peisl, F. Grey, R.L. Johnson and S. Krummacher, Phys. Rev. Lett. **60**, 2382 (1988)

[3] K. Al Usta, H. Dosch, A. Lied and J. Peisl, accepted in Physica B (1991)

[4] S. Dietrich and H. Wagner, Z. Phys. B 59, 35 (1985)

[5] H. Dosch, Phys. Rev. B35, 2137 (1987)

[6] A. Zeilinger and T.J. Beatty, Phys. Rev. B 27, 7239 (1983)

[7] J.F Ankner, H. Zabel, D.A. Neumann and C.F. Majkrzak, Phys. Rev. B 40, 792 (1989)

[8] P Heller and G.B. Benedek, Phys. Rev. Lett. 8, 428 (1962)

[9] R.A. Erickson, Phys. Rev. 90, 779 (1953)

[10] K Al Usta, H. Dosch, A. Lied and J. Peisl, to be published

Analytical Calculation of the Resolution Correction Function for X-Ray Surface Structure Analysis at High Exit Angles

C. Schamper, H.L. Meyerheim, W. Moritz, and H. Schulz

Institute of Crystallography and Mineralogy, University of Munich, Theresienstr. 41, W-8000 München 2, Fed.Rep.of Germany

Abstract. The analysis of rod scans in surface x-ray diffraction provides information about structure parameters normal to the sample surface. In order to achieve high resolution the measurements have to extend to momentum transfers q_z as large as possible. The proper correction of the measured intensity data for the resolution function of the detector is a prerequisite for obtaining reliable structure information. We have developed an analytical expression for the resolution correction of rod scan intensity data which take into account an anisotropic detector resolution $T(\Theta,\Phi)$, the domain size of the sample, ζ, and the primary beam divergence parallel to the sample surface $\Delta\tau$.

1. Introduction

Glancing incidence angle x-ray scattering is now a well established technique for surface structure analysis [1,2]. So far work has been concentrated on measurements of 'in plane' data providing only the projection of the structure. In order to obtain infomation about the structural parameters normal to the sample surface, measurments along the reciprocal lattice rods up to high momentum transfer $q_z=\ell\cdot c^*$ have to be performed. However, in all diffraction experiments the measured intensities have to be corrected for the instrumental resolution. For in plane structure analysis this is simply done by multiplying the intensity by a factor Δq_z corresponding to the length of the rod accepted by the detector [1,2].

The correction factors become complicated for rod scans, when the scattering plane is no longer parallel to the sample surface. In the following we shortly outline the intensity correction due to the detector resolution for our z-axis diffractometer with two independent detector angles. Subsequently we discuss the results by considering different detector arrangements, where the primary beam divergence and the finite domain size of the sample are taken into account.

2. Calculation

Experimentally, the integrated intensity is measured by rotating the sample around the surface normal keeping the detector at a fixed position out of plane. In Fig.1 this procedure is shown in projection to the $\ell=0$ plane when the reciprocal lattice rod is swept through the detector's aperture. The integrated intensity is given by the convolution of the reflectivity with the transmission function:

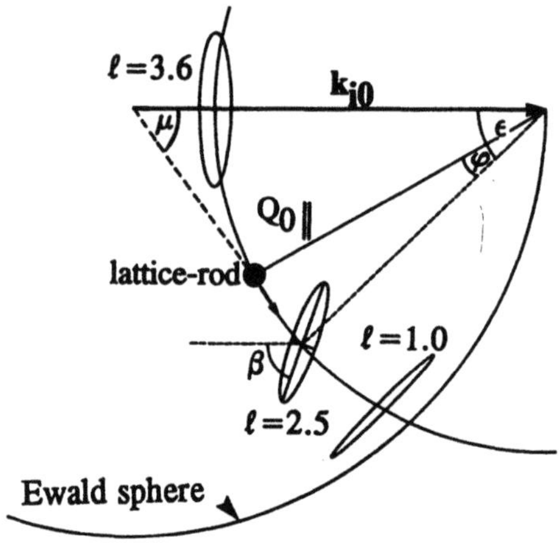

Fig. 1: Projection of the diffraction geometry on the $\ell=0$ plane, illustrating the measurement of the integrated intensity. The sample is rotated around the surface normal. Diffraction is obtained when the rod intersects the detector aperture. The angular range of φ where the rod remains within the detector changes at different ℓ values. The measured intensities must therefore be corrected due to the different shapes of the resolution ellipse.

$$E(Q_0) \propto \iiint i(k_i-k_{i0}) \cdot |F(q)|^2 \cdot$$
$$\cdot\, T(q-Q_0-\Delta Q_1(k_i-k_{i0})-\Delta Q_2(\varphi))\, dq^3\, dk_i^3\, d\varphi \quad . \tag{1}$$

In Eq. 1 $i(k_i-k_{i0})$ describes the intensity distribution in the primary beam around the incident beam k_{i0}. Gaussian functions are assumed for the primary beam intensity distribution i, the transmission function T and the rod profile parallel to the surface $|F(q_\parallel)|^2$. The parameters q and Q_0 represent vectors in reciprocal space, the latter indicates the reflection under consideration. The primary beam divergence and the rotation angle φ of the crystal are taken into account by the arguments $\Delta Q_1(k_i-k_{i0})$ and $\Delta Q_2(\varphi)$. The calulation is based on three assumptions: (1) a grazing incidence geometry is used, (2) the radiation is strictly monochromatic and (3) the lateral extensions of the reciprocal lattice rod and the detector acceptance are small as compared to the radius of the Ewald sphere. The analytic solution of Eq. (1) is simple but tedious and will be described elsewhere [3]. Finally we are left with an expression $E(Q_0) \propto |F(Q_0)|^2 \cdot L(Q_0)$, where

$$L(Q_0) = \frac{\lambda \cdot \Delta\Theta \cdot \Delta\Phi}{d^{1/2} \cdot a \cdot b \cdot |Q_{0\parallel}|} \left\{ 1 + m \cdot \Delta\tau^2 \cdot \Gamma^2 + m/\zeta^2 \right\}^{-1/2} \tag{2}$$

indicates the Lorentz factor which describes the dependence of the integrated intensity on pure geometric parameters, particularly on the acceptance angles of the detector $\Delta\Theta$ and $\Delta\Phi$. The parameters d, m and Γ are given by

$$c = \cos(\beta-\epsilon) \cdot \cos(\beta-\epsilon) \cdot (1/a^2 - 1/b^2) \tag{3}$$
$$d = (\sin(\beta-\epsilon))^2 / a^2 + (\cos(\beta-\epsilon))^2 / b^2 \tag{4}$$
$$e = (\cos(\beta-\epsilon))^2 / a^2 + (\sin(\beta-\epsilon))^2 / b^2 \tag{5}$$

$$m = \lambda^2 \cdot (e - c^2/d) \qquad [\text{Å}^2] \qquad (6)$$

$$\Gamma = (\cos\mu)/\lambda - |Q_{0\|}| \cdot \cos(\mu+\epsilon), \qquad [\text{Å}^{-1}] \qquad (7)$$

where a/λ and b/λ [Å^{-1}] are the short and the long axis of the projected resolution ellipse, respectively (Fig. 1) and μ is the projection of the scattering angle to the surface. The angles β and ϵ are explained in Fig. 1. The Lorentz factor factor given by Eq. (2) is composed of two terms, the first representing the intensity correction assuming δ-function like lattice rods and strictly parallel incident radiation, the second (under the square root) takes account for the domain size, ζ, and the primary beam divergence $\Delta\tau$. In the low exit angle regime ($\ell \leq 1$) the correction calculated from Eq. (1) agrees with the published formula [1,2], provided that the domain size ζ is large and the primary beam is parallel ($\Delta\tau=0$).

3. Results and Discussion

In order to elucidate the different factors contributing to the intensity corrections we have calculated $L(\ell)$ for the (1/2 1/2) and (3/2 5/2) superlattice reflection rod of the Ge(001)(2x1) structure assuming a wavelength of λ=1.54 Å. The results are shown in Fig. 2. In order to maximize the intensity at low ℓ it is evident that the detector has to be aligned with a high acceptance parallel to the rods. However, the situation changes at intermediate and high ℓ, where the intersection of the lattice rod with the Ewald sphere crosses the detector transmission increasingly oblique to the long axis. Under these conditions it is preferable to rotate the detector by 90°, as displayed in Fig.1. Additionally, the dependence on $Q_{0\|}$ results in an about four times larger intensity of the (1/2 1/2 ℓ) as compared with the (3/2 5/2 ℓ) reflections for $\ell \leq 1$.

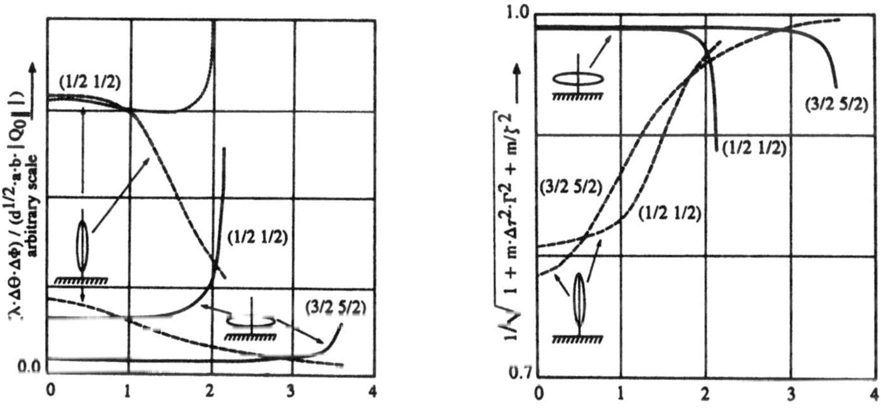

Fig. 2: Lorentz factor as a function of ℓ for both detector settings (solid and dashed line) assuming parallel primary beam and δ-like rods (left panel). The detector acceptance angles are 0.4° and 2.0°. The dotted line indicates the resolution correction for the (1/2 1/2) rod as given by ref. [2]. The right panel shows the ℓ-dependence of the correction term in Eq. 2 for a domain size ζ of 500 Å and a beam divergence $\Delta\tau$ of 0.23°.

In the right panel of Fig. 2 we have plotted the square root expression in Eq. 2 versus ℓ including a primary beam divergence of $\Delta\tau$=0.23° and a domain size of ζ=500 Å corresponding to our beam collimation and a good quality sample, respectively. This illustrates the effect of realistic beam and sample conditions leading to a broadening of the reflection profile. The broadening of the lattice rods does not have an influence at low ℓ if the detector transmission function is aligned with the short axis parallel to the rods as indicated. In contrast a 10 to 20% intensity decrease is observed if the long axis is aligned parallel to the lattice rods (e.g. high resolution in plane), since the detector does not accept the full width of the reflection. This result is of considerable importance for measurements of 'out of plane' data with large beam divergence using a rotating anode and imposes stringent demands on the accuracy of the detector alignment.

Acknowlegement. This work was supported by the Bundesminister für Forschung und Technologie Grant No. 05464IAB8.

References

[1] I.K. Robinson, in: Handbook on Synchrotron Radiation, Vol. 3, Eds. D.E. Moncton and G.S. Brown (North Holland, Amsterdam, 1990)
[2] R. Feidenhans'l, Surf. Sci. Rep. 10, 105, (1989)
[3] C. Schamper, H.L. Meyerheim, W. Moritz, to be published

Neutron Double Crystal Diffractometry – A Precise Method for Surface Investigations

F. Eichhorn[1], *K.M. Podurets*[2], *S.Sh. Shilstein*[2], and *Z.N. Soroko*[3]

[1]Central Institute for Nuclear Research Rossendorf, P.O. Box 19, O-8051 Dresden, Fed. Rep. of Germany

[2]I.V. Kurchatov Institute of Atomic Energy, SU-123182 Moscow, USSR

[3]Leningrad Nuclear Physics Institute, SU-188350 Gatchina, USSR

Abstract. To investigate the surface reflectivity a neutron double crystal diffractometer with perfect crystals can be used. By this technique it is possible to measure the reflectivity and the geometrical perfection of surfaces independently. The refractive index can be determined within the 0.1 percent limit and the surface flatness with an accuracy of several seconds of arc.

Total reflection measurements give information about the neutron refractive index. The critical angle $\theta_c = \lambda\sqrt{N\bar{b}/\pi}$ is of the order of 10^{-3} for neutrons where λ is the wavelength, N is the number of nuclei per volume, and $\bar{b}$ is the mean coherent scattering amplitude. Most neutron reflectivity measurements are limited by the angular ($\delta\theta$) and wavelength spread ($\delta\lambda$). A way to avoid this difficulty is to use a double-crystal diffractometer with perfect crystals.

If a polychromatic and divergent beam incidents on a perfect crystal then within the diffracted beam a correlation between wavelength λ and propagation direction θ exists in the limit of the diffraction curve width (seconds of arc) of the dynamical theory. Such a beam falls on a flat reflecting surface as seen in fig. 1. One part of the neutron beam (beam 1 in fig. 1a and experimental results in fig. 2a) does not interact with the mirror and is diffracted by the second crystal if that is parallel to the first one what gives the experimental origin for the β axis. The second part is the beam refracted by the mirror (beam 2). A third part (beam 3) is incident on the mirror at $\theta \approx \theta_c$ and is reflected at its surface.

In the (+,–) position of the diffractometer the angle of the analyzer crystal at the total reflection jump is given by [1] $\beta_{+-c} = 2\theta_c \cdot [1 + (\theta_c - \alpha)/\tan\theta_B]$, here α is the angle between the central beam and the mirror surface and θ_B is the Bragg angle of the diffractometer crystals. Therefore, in the case of critical reflection the position β_{+-c} of the intensity jump depends only weakly on the mirror orientation, $\Delta\beta_{+-c} \approx 10^{-3} \cdot \Delta\alpha$.

In the (+,+) setting of the diffractometer crystals (fig. 1b and 2b) the specularly reflected beam has the angular width $\delta\beta_{++} =$

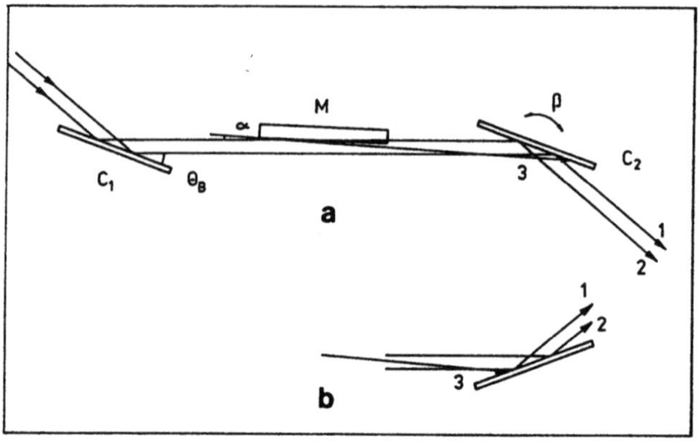

Fig. 1. Double-crystal diffractometer in the (+,−) (a) and (+,+) (b) setting. 1 – beam diffracted only by the crystals, 2 – beam refracted by the mirror material, 3 – beam reflected at the mirror surface, C_1 and C_2 – diffractometer crystals, M – mirror.

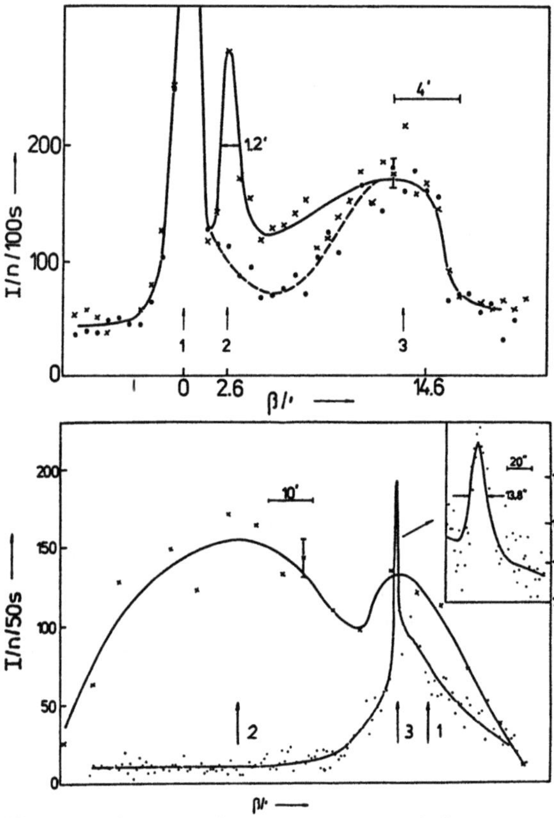

Fig. 2. Neutron beam intensity diffracted at the second crystal versus its rotation angle in the (+,−) (a) and (+,+) (b) setting of the crystals for a glass mirror ($\lambda = (0.202 \pm 0.001)$ nm). Beams 1, 2, and 3 like in fig. 1. (∘ – mirror end faces are covered with Cd.)

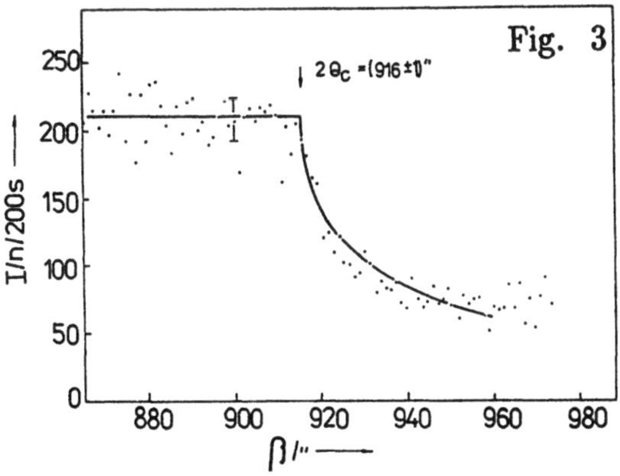

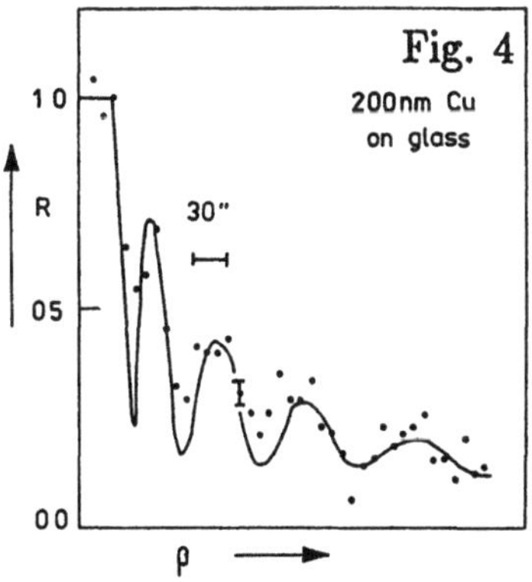

Fig. 3. Intensity of the reflected beam in the vicinity of the critical angle for a glass mirror. The solid line was calculated.

Fig. 4. Intensity of the beam specularly reflected by a 200 nm thick Cu film on glass.

$2\delta\alpha$, and is determined by the geometrical perfection of the surface.

The objects of measurement were plates of glass used for neutron guides. Results are shown in fig. 2 and 3. The possibility of investigating films was demonstrated with a Cu film of 200 nm thickness on glass (fig. 4). Due to the high angular resolution of the double-crystal arrangement films of a thickness up to 500 nm can be studied.

The double-crystal diffractometer with perfect crystals allows to measure the critical angle of total external reflection with an accuracy of 1". This corresponds to an accuracy of 10^{-3} in the refractive index determination. The precision of these values are nearly independent of the surface curvature. This technique allows to carry out simultaneous and separate investigations of material properties (refractive index) as well as of the geometrical perfection of surfaces with a high precision.

References

[1] K. M. Podurets, S. Sh. Shilstein, F. Eichhorn, Nucl. Instr. and Meth. **A 290**, 419 (1990).

Index of Contributors

Springer Proceedings in Physics

Managing Editor: H. K. V. Lotsch

1 *Fluctuations and Sensitivity in Nonequilibrium Systems*
Editors: W. Horsthemke and D. K. Kondepudi

2 *EXAFS and Near Edge Structure III*
Editors: K. O. Hodgson, B. Hedman, and J. E. Penner-Hahn

3 *Nonlinear Phenomena in Physics* Editor: F. Claro

4 *Time-Resolved Vibrational Spectroscopy*
Editors: A. Laubereau and M. Stockburger

5 *Physics of Finely Divided Matter*
Editors: N. Boccara and M. Daoud

6 *Aerogels* Editor: J. Fricke

7 *Nonlinear Optics: Materials and Devices*
Editors: C. Flytzanis and J. L. Oudar

8 *Optical Bistability III*
Editors: H. M. Gibbs, P. Mandel, N. Peyghambarian, and S. D. Smith

9 *Ion Formation from Organic Solids (IFOS III)*
Editor: A. Benninghoven

10 *Atomic Transport and Defects in Metals by Neutron Scattering*
Editors: C. Janot, W. Petry, D. Richter, and T. Springer

11 *Biophysical Effects of Steady Magnetic Fields*
Editors: G. Maret, J. Kiepenheuer, and N. Boccara

12 *Quantum Optics IV*
Editors: J. D. Harvey and D. F. Walls

13 *The Physics and Fabrication of Microstructures and Microdevices*
Editors: M. J. Kelly and C. Weisbuch

14 *Magnetic Properties of Low-Dimensional Systems*
Editors: L. M. Falicov and J. L. Morán-López

15 *Gas Flow and Chemical Lasers*
Editor: S. Rosenwaks

16 *Photons and Continuum States of Atoms and Molecules*
Editors: N. K. Rahman, C. Guidotti, and M. Allegrini

17 *Quantum Aspects of Molecular Motions in Solids*
Editors: A. Heidemann, A. Magerl, M. Prager, D. Richter, and T. Springer

18 *Electro-optic and Photorefractive Materials*
Editor: P. Günter

19 *Lasers and Synergetics*
Editors: R. Graham and A. Wunderlin

20 *Primary Processes in Photobiology*
Editor: T. Kobayashi

21 *Physics of Amphiphilic Layers*
Editors: J. Meunier, D. Langevin, and N. Boccara

22 *Semiconductor Interfaces: Formation and Properties*
Editors: G. Le Lay, J. Derrien, and N. Boccara

23 *Magnetic Excitations and Fluctuations II*
Editors: U. Balucani, S. W. Lovesey, M. G. Rasetti, and V. Tognetti

24 *Recent Topics in Theoretical Physics*
Editor: H. Takayama

25 *Excitons in Confined Systems*
Editors: R. Del Sole, A. D'Andrea, and A. Lapiccirella

26 *The Elementary Structure of Matter*
Editors: J.-M. Richard, E. Aslanides, and N. Boccara

27 *Competing Interactions and Microstructures: Statics and Dynamics*
Editors: R. LeSar, A. Bishop, and R. Heffner

28 *Anderson Localization*
Editors: T. Ando and H. Fukuyama

29 *Polymer Motion in Dense Systems*
Editors: D. Richter and T. Springer

30 *Short-Wavelength Lasers and Their Applications*
Editor: C. Yamanaka

31 *Quantum String Theory*
Editors: N. Kawamoto and T. Kugo

32 *Universalities in Condensed Matter*
Editors: R. Jullien, L. Peliti, R. Rammal, and N. Boccara

33 *Computer Simulation Studies in Condensed Matter Physics: Recent Developments*
Editors: D. P. Landau, K. K. Mon, and H.-B. Schüttler

34 *Amorphous and Crystalline Silicon Carbide and Related Materials*
Editors: G. L. Harris and C. Y.-W. Yang

35 *Polycrystalline Semiconductors: Grain Boundaries and Interfaces*
Editors: H. J. Möller, H. P. Strunk, and J. H. Werner

36 *Nonlinear Optics of Organics and Semiconductors*
Editors: T. Kobayashi

37 *Dynamics of Disordered Materials*
Editors: D. Richter, A. J. Dianoux, W. Petry, and J. Teixeira

38 *Electroluminescence*
Editors: S. Shionoya and H. Kobayashi

39 *Disorder and Nonlinearity*
Editors: A. R. Bishop, D. K. Campbell, and S. Pnevmatikos

40 *Static and Dynamic Properties of Liquids*
Editors: M. Davidović and A. K. Soper

41 *Quantum Optics V*
Editors: J. D. Harvey and D. F. Walls

42 *Molecular Basis of Polymer Networks*
Editors: A. Baumgärtner and C. E. Picot

43 *Amorphous and Crystalline Silicon Carbide II: Recent Developments*
Editors: M. M. Rahman, C. Y.-W. Yang, and G. L. Harris

FSC
www.fsc.org
MIX
Papier aus verantwortungsvollen Quellen
Paper from responsible sources
FSC® C105338

If you have any concerns about our products, you can contact us on
ProductSafety@springernature.com

In case Publisher is established outside the EU, the EU authorized representative is:
Springer Nature Customer Service Center GmbH
Europaplatz 3, 69115 Heidelberg, Germany

Printed by Libri Plureos GmbH
in Hamburg, Germany